Microfarms

Organic Market Gardening on a Human Scale

JEAN-MARTIN FORTIER
AND AURÉLIE SÉCHERET

Translated by Laurie Bennett

new society
PUBLISHERS

Cover design by Diane McIntosh.
Cover image credits: Top image, bottom right, bottom left © Anne-Claire Héraud;
middle bottom and center © Market Gardener Institute.
For full interior photo credits see page 150.
Printed in Canada. First printing December, 2024.

© Delachaux et Niestlé, Paris, 2023
First published in France under the title:
Microfermes. Le maraîchage bio à l'échelle humaine
Jean-Martin Fortier, Aurélie Sécheret

This book is intended to be educational and informative. It is not intended to serve as a guide. The author and publisher disclaim all responsibility for any liability, loss, or risk that may be associated with the application of any of the contents of this book.

Inquiries regarding requests to reprint all or part of *Microfarms* should be addressed to New Society Publishers at the address below. To order directly from the publishers, please call 250-247-9737 or order online at www.newsociety.com.

Any other inquiries can be directed by mail to:
New Society Publishers P.O. Box 189, Gabriola Island, BC V0R 1X0, Canada

Library and Archives Canada Cataloguing in Publication

Title: Microfarms : organic market gardening on a human scale / Jean-Martin Fortier and Aurélie Sécheret ; translated by Laurie Bennett.

Other titles: Microfermes. English

Names: Fortier, Jean-Martin, author | Sécheret, Aurélie, author. | Bennett, Laurie, translator.

Description: Translation of: Microfermes: le maraîchage bio à l'échelle humaine. | Includes bibliographical references.

Identifiers: Canadiana (print) 20240441486 | Canadiana (ebook) 20240441494 | ISBN 9781774060049 (softcover) | ISBN 9781550927979 (EPUB) | ISBN 9781771423939 (PDF)

Subjects: LCSH: Farms, Small—Case studies. | LCSH: Sustainable agriculture. | LCSH: Organic farming. | LCSH: Truck farming. | LCGFT: Case studies.

Classification: LCC S494.5.S86 F67 2024 | DDC 631.58—dc23

Funded by the Government of Canada
Financé par le gouvernement du Canada

Canada

Contents

Publisher's Foreword

As I read the English translation of this manuscript, I became increasingly excited at the diverse range of farmers being profiled, each with a unique backstory and trajectory. I felt like I had just unearthed a treasure chest of possibilities for my own microfarm, and I immediately wanted to share this bounty with my neighbors who are also engaged in the small-scale farming revolution launched more than a decade ago by Jean-Martin Fortier with the landmark book *The Market Gardener*.

When I told Jean-Martin that I was keen to publish the book in English, he asked me point-blank why I thought a collection of profiles of small-scale farms located largely in Europe would be of interest to the small-scale farming communities across the United States, Canada, and the UK. My answer to him, and the reason for publishing this book in English, is that it's a testament to the power of diversity.

As someone who lived abroad for a decade and spent years circling the globe to dozens of countries, I've learned firsthand that diverse places and cultures do many things—including farming—differently than at home, wherever home may be. And "different" doesn't mean, as we so often default to in our cultural silos, "better" or "worse." Rather, "different" means an expansion of mind and worldview, a dissolving of the certainty that "my way is the right way," and a promotion of both inquiry and, vitally, the humility that others, elsewhere, who speak different languages and come from unique cultural perspectives, have vital knowledge to share if we are willing to listen.

It's with this embrace of the power of "different" that I present this collection to you. Each case study—be it a peri-urban microfarm, a farm built on a seasonal floodplain, or a market garden tucked between the rows of an ancient vineyard—highlights the unique contexts, challenges, and triumphs of microfarmers. This diversity is where the power lies in our collective cultural knowledge and wisdom, forged over time, through trial and error, in a range of circumstances.

As you will see in these pages, there is no single portrait of a farmer, nor one way to farm; rather, it is a mosaic of people, places, and possibilities. I encourage you to embrace the "different," and open your mind and heart to what is possible as we all work together to co-create the ongoing small-scale farming revolution.

Happy growing!

Rob West, Acquisitions Editor
New Society Publishers
September 2024

Preface

Much has been written about the organic microfarming concept. While the ecological benefits of this approach to farming are now proven, financial returns have not always been as good. The reason for this is simple: the people spearheading microfarming projects were either environmental activists and dreamers or wealthy businessmen—and none of them had a strong grasp of both growing techniques and business models. This led to many failures. It's important to recognize that microfarming—working with approximately 2.5 acres—requires a complex set of skills and knowledge. Complex, yes, but nowhere near impossible.

The successes of the eight farms presented in this book are based on methods that, while sometimes tweaked on a case-by-case basis, leave no room for improvisation. Jean-Martin Fortier spent years learning in the field to develop the agricultural techniques that are at the heart of this method for biointensive systems. For microfarms, this method is the first pillar of success. The second pillar is, of course, having a clear picture of the farm's business model. The model requires a few years of operation to generate modest profits, and must be managed perfectly. The same is true for investments, which you need to value accurately. The third and final pillar is distribution. Here, efficiency and speed are critical.

If you can get this equation right—as we did at Ferme du Perche—you are sure to run a profitable business. For us, that meant €300,000 (US$325,000) in investments, a €100,000 (US$108,000) loss in the first two years, and now €250,000 (US$270,000) in revenue, with six full-time people. At Ferme du Perche, we became profitable by amortizing our assets, and we did this without public grants or subsidies as they were not suited to our specific microfarming system. Profitability could improve if grant reforms were to extend coverage to small-scale operations and compensate farmers for the ecological services they provide.

The prospect of resurrecting market gardening techniques from the nineteenth century, of serving the planet and humans by growing healthy fruits and vegetables, but also seeing your operations become sustainable… There are so many good reasons to start a farm. So don't wait any longer!

Jean-François Rial,
P. D.G. de Voyageurs du Monde,
cofondateur d'Une Ferme du Perche

The Fortier Method for a Successful Microfarm

Farming on a Human Scale

Jean-Martin Fortier's biointensive farming model integrates principles from agroecology, permaculture, and entrepreneurship, emphasizing a nonmechanized approach he terms as "farming on a human scale." At its core, "farming on a human scale" rejects reliance on machinery. It empowers organic growers to manage their businesses autonomously, emphasizing crop diversity, multi-cropping, and direct-to-customer sales and marketing. This approach supports sustainable livelihoods for small landholders in the developed world. It embraces low-tech, cost-effective technologies that prioritize the expertise, skills, and craftsmanship of those tending the fields.

Jean-Martin's journey began at the McGill School of Environment in Montréal, where he met his wife and collaborator, Maude-Hélène Desroches. Motivated by a shared vision for positive impact, they gained practical experience on an organic farm in New Mexico, USA. They also visited Cuba's innovative urban agricultural models, developed during the country's Special Period, when it faced economic isolation. There, they saw firsthand how organic agriculture, free from chemical fertilizers, pesticides, herbicides, and tractors, could sustainably feed a population.

Upon their return to Quebec in 2004, Jean-Martin and Maude-Hélène purchased a modest 10-acre plot in Saint-Armand, a picturesque farming village in Quebec's Eastern Townships. They named their farm Les Jardins de la Grelinette, inspired by the *grelinette* (the French word for "broadfork"), a tool symbolizing their commitment to no-till farming and soil ecology. With only two acres suitable for cultivation, they focused on optimizing production. They were initially guided by Eliot Coleman's organic farming principles, which prioritize densely planted permanent beds over traditional row cultivation, and they shaped the farm's design around this philosophy. In every aspect of their farming operation, they integrated efficiency by embracing innovative hand tools that predominantly hailed from Europe and were designed for ergonomic efficiency, much like larger mechanized machinery. Their initial tool kit included push seeders, hoes, a two-wheel walk-behind tractor, and other essential implements.

Drawing from the work of professional greenhouse growers, they quickly mastered the art of greenhouse crop cultivation, which significantly increased yields in their most profitable crops. Within the first few years, crop rotation, integrated pest management, and optimal fertility practices became the cornerstones of their farming system. Leveraging low-tech season extension methods alongside a sophisticated crop planning model, they achieved 3 to 4 crop rotations per year in a given bed, despite the challenging northeastern climate.

Les Jardins de la Grelinette exclusively marketed their produce through direct channels like CSA boxes, two farmers' markets, and local grocery stores, where they sold their famous mesclun. This approach allowed them to bypass intermediaries and ensure maximum returns on sales.

Making Less than Two Acres Profitable

By about 2008, Jean-Martin and Maude-Hélène had already achieved swift success with their farm. They generated Can$43,675 in their first year, doubled their earnings the following year, and exceeded Can$145,800 by the third year. By the fourth and fifth years, sales of produce from the farm were close to Can$200,000.

This early success sustained their family and also garnered recognition from the Quebec Department of Agriculture, which, through ongoing financial programs for farmers, has acknowledged the benefits that can be generated on only a few acres of land.

Over the next four years, the couple grew the farm without expanding their land base, increasing yields and sales every year. In 2012, Jean-Martin decided to share their findings in a book titled *The Market Gardener*, a comprehensive manual detailing their farming model and strategies. The book became an instant bestseller, eventually selling over 250,000 copies worldwide, and it has been translated into ten languages.

In 2015, following an invitation from his mentor Eliot Coleman, Jean-Martin embarked on a new project and established La Ferme des Quatre-Temps, a research and education farm funded by a wealthy Canadian family. The goal of this initiative was to reimagine the future of farming, making it more holistic and regenerative. At its core they set

Located in Hemmingford, Quebec, and less than an hour from Montréal, La Ferme des Quatre-Temps spans 160 acres and features pasture-raised livestock, a culinary lab, and 7.5 acres of market gardens. Every year, it takes on ten apprentice market gardeners who are taught the principles developed at the farm. For several seasons, a film crew followed Jean-Martin. This resulted in the popular TV show *Les fermiers*, which aired for two seasons and is available on TV5 Monde and Apple TV.

up a training ground that became Jean-Martin's farming school. Over the next seven years, he annually trained a cohort of ten young farmers, applying the Jardins de la Grelinette methods, while also pioneering new strategies and techniques to optimize productivity on a human scale. These ideas and strategies were eventually encapsulated by the term "biointensive," which had previously been used by innovators in California's back-to-the-land movement in the 1960s.

Continuing his mission to educate and inspire people, Jean-Martin then launched an online course called the Market Gardener Masterclass, featuring his market gardening methodologies and bringing them to a wider audience. Now available in more than 90 countries, the masterclass teachings are the cornerstone of the Market Gardener Institute, which aims to help improve the skills and knowledge of small farms around the globe.

A Model for Hope

Small-scale organic farms, or microfarms, have the potential to feed the world, and Jean-Martin and Maude-Hélène are shining examples of this. Their model not only demonstrates the profitability of bio-intensive market gardening strategies but also their ability to yield five times more produce per unit of land when compared to conventional farming. Moreover, their approach emphasizes sustainability, which is underpinned by Maude-Hélène's steadfast commitment to their farming lifestyle. As Les Jardins de la Grelinette enters its twentieth season of production, it continues to deliver high-quality, high-volume crops, proving the long-term viability of these methods.

Their microfarm, along with thousands of others directly influenced by their model, now serves as a blueprint for aspiring farmers, regardless of their agricultural background. The model not only promotes a low-tech, low-investment, high-return approach but is also an inspiration for individuals looking to start their farming journey, as it has shown that long-term financial independence is possible.

As the examples in this book demonstrate, their approach has been replicated around the world. Their model for small-scale agriculture seems to fit wherever there is a strong demand for local food and systems that allow farmers to avoid agricultural scaling mechanisms, which tend to result in diminishing returns on sales and increased complexity. This contrast is best described as quality and client relationships over volume.

The Future Might Be In the Past

Principles of the Method

The principles of the Les Jardins de la Grelinette model described in *The Market Gardener* are not new. They are rooted in a rich historical farming tradition. In the nineteenth century, Parisian market gardeners fed a city of over two million people through thousands of microfarms located within the city limits. By the 1830s, it is estimated that more than 10,000 small plots supplied fresh vegetables to city inhabitants who, in turn, provided horse manure that sustained soil fertility. The model was praised for its productivity and closed-loop sustainability.*

These market gardeners employed organic, nonmechanized techniques to meet urban food demands amidst rapid urbanization, showcasing remarkable ingenuity. Throughout the nineteenth and twentieth centuries, their methods and technical achievements were celebrated in literature across Europe. However, with the rise of modern agricultural practices, much of this knowledge faded into obscurity, even in France. Mechanization, modern agronomic science, and improved transportation of inexpensive foreign produce shifted small farms towards larger, less diverse, and more technologically driven operations—a trend that continues to shape agriculture today.

Nevertheless, the Parisian market gardeners' horticultural methods laid a foundation for enduring sustainable farming practices. These methods cleverly maximize yields without degrading soil, all while operating on a human scale with minimal mechanization. As this book illustrates, the techniques are profitable in a modern context, mainly because they rely on only modest start-up investments and require a smaller workforce. Both factors make it easier to become profitable quickly.

The success of the farms presented in this book exemplifies how efficient processes can flourish on small plots of land used for intensive agriculture. Still, the path forward is not easy; for those who are committed to the farming lifestyle, it requires dedication and hard work.

Good Design and Planning: Essentials

The biointensive model focuses on maximizing yields per square foot (or square meter). However, achieving higher yields isn't just about planting more densely; it requires an extensive array of skills and techniques that can be developed over time. Jean-Martin Fortier has shared these insights with organic growers worldwide, highlighting the importance of building a well-rounded range of expertise. This book aims to share some of these insights in a structured manner, offering guidance on how to effectively implement and benefit from biointensive farming practices.

Proper farm design lays the foundations for maximum efficiency and sustainable success. In the initial stages, farmers must tackle intensive tasks like preparing fields, building greenhouses, installing

* J.G. Moreau and J.J. Daverne, *Manuel pratique de la culture maraîchère de Paris* (Paris, 1845), available free and in French only, at gallica.bnf.fr

irrigation systems, and setting up essential infrastructure. To improve overall efficiency on the farm, these systems need to work together seamlessly—this is a central principle of permaculture.

Effective farm design encompasses various crucial elements beyond infrastructure setup. This includes meticulously sizing bed lengths, standardizing the overall layout, storing tools in strategic locations, making sure that counters in the washing and packing station are at the right height, appropriately sizing cold rooms, and optimizing other systems.

Understanding the farm layout from a bird's-eye view provides invaluable insights into its functionality and efficiency. Throughout this book, we present diverse farm designs that exemplify these principles, illustrating how thoughtful planning and integration can significantly boost the efficiency of small-scale farming operations.

Crop planning is another cornerstone of successful small-scale farming and market gardening. On small farms, the myriad tasks can

quickly become overwhelming, especially when attempting to double, triple, or quadruple the number of crops grown in a given bed, which is a key tenet of the biointensive approach taught by Jean-Martin. Growers can achieve success by mastering the art of crop management through careful planning.

To optimize time and space in the garden, growers need to be strategic with crop successions, in a process that involves planting a new crop in a bed as soon as the last crop has been removed. As described in *The Market Gardener*, seamless transitions between crops require meticulous planning, guided by five main questions:

- Which vegetables should you grow?
- In what quantities?
- When should you harvest them?
- When should you start and transplant all your seedlings?
- Where should you plant these crops so it will be easier to time crop successions?

By addressing these questions and organizing calendars that outline tasks, growers can learn to streamline chores and, most importantly, alleviate the mental load associated with a complex crop planting schedule. This structured approach not only enhances efficiency but also ensures that the use of each bed is optimized throughout the growing season. It allows farmers to focus their efforts on timely planting, crop maintenance, and harvests, thereby maximizing the yields of their small-scale operation.

Steps
to
Success

THE MARKET GARDENER'S STEPS TO SUCCESSFUL CROP PLANNING

Step 1: Establish your goals and financial targets for the season

1. **Review your mission statement:** Begin by revisiting your farm's mission statement. Understand why you are in this line of work and what values drive your farm's operations. This will help align your goals with your broader purpose and vision.
2. **Set professional goals:** Define specific professional goals for the upcoming season. Consider aspects such as expanding market reach, improving crop yields, adopting new sustainable practices, or enhancing customer satisfaction.
3. **Reflect on your life goals:** Beyond professional aspirations, reflect on your personal life goals. How does farming contribute to your overall life satisfaction and fulfillment? Understanding this connection can motivate and guide your decisions throughout the season.
4. **Establish financial targets:** Outline clear financial targets for the year. Calculate expected revenue and set realistic profit goals based on market trends, production capabilities, and operational costs.

By setting clear goals and financial targets from the outset, you are establishing a framework for decision-making and strategic planning throughout the farming season. This clarity will help prioritize actions that align with both your farm's mission and your personal aspirations.

Step 2: Make a production plan

1. **Determine your points of sale:** Choose your sales outlets based on your location, preferences, and market demand. Consider options like restaurants, CSA boxes, farmers' markets, and on-farm sales. Avoid relying solely on one outlet to mitigate risk and have diversified revenue streams.
2. **Decide which vegetables to grow:** Select between 30 and 40 crops that align with market demand and your farm's capabilities. Focus on high-value crops that customers want, balancing yield potential with taste and uniqueness. Avoid crops that are labor-intensive or yield less income per square foot.
3. **Calculate time and bed space for each crop:** Estimate the yield per bed for each chosen crop and figure out which planting and harvest dates will align with your sales calendar. Plan multiple successions for crops like carrots to ensure a continuous supply throughout the season.

Example using carrots:
Spring
- **Begin in January:** Seed 4 beds of early carrots in the tomato greenhouse, to harvest before planting tomatoes.
- **Mid-March:** Seed another 4 beds in caterpillar tunnels for early season demand.

Summer
- **April 15 to August 8:** Seed 2 beds per week in the field to meet weekly demand (approx. 360 bunches per week).

Fall
- **August 8:** Seed 6 beds of storage carrots in the field for the winter.

4. **Document your plan:** Compile all crop details into a single document for easy reference. Include the crop type, planting dates, expected yields, and bed requirements. At this stage, you need to assign crops to specific growing spaces like fields, tunnels, or greenhouses, but you do not need to specify which bed will receive each one.

A comprehensive production plan will help make sure you meet market demands efficiently, optimize bed space, and maintain a steady supply of high-quality produce throughout the growing season. Use this plan as a guide to seamlessly coordinate planting, harvesting, and sales.

Step 3: Map out the farm

Creating a detailed crop map is crucial for visualizing and managing your farm's operations effectively throughout the growing season.

1. **Choose your tool:** Use Microsoft Excel or dedicated crop planning software to create your crop map. Alternatively, you can use paper and pen for a manual approach.
2. **Draw out field blocks:** In Excel, draw rectangles to represent each bed or growing area. Label each rectangle with the crop name, succession number (1st, 2nd, etc.), planting dates, and harvest dates.
3. **Optimize space:** To get the most out of each bed, be strategic when choosing where to plant crops. Start with high-demand crops. Limit downtime between successions by covering beds with tarps for 2-week intervals to prepare them for the next crop.

4. **Organize by crop needs:** Group crops with similar watering needs together to streamline your irrigation setup. This will allow you to use drip or sprinkler systems efficiently across these combined beds.

5. **Maintain clean beds:** Before planting, cover beds with tarps to minimize debris and ensure a clean planting environment. This practice helps reduce weed growth and maintain soil health.

6. **Implement crop rotations:** Follow a crop rotation plan to mitigate damage caused by pests, disease, and nutrient depletion. Avoid planting consecutive crops from the same family in one area whenever possible.

7. **Adjust for space constraints:** If you find you don't have enough total bed space to accommodate your production plan, prioritize high-value crops or create new blocks to meet the demand. Ensure your production plan, financial targets, and crop map are aligned and coherent.

8. **Time commitment:** Creating a comprehensive crop map may take 2 to 7 days, depending on the complexity of your farm and the tools used. For thorough and accurate planning, dedicate focused time to tackle this with your team.

Creating a detailed map of your farm will help enhance decision-making, optimize the use of resources, and streamline operations throughout the growing season. This visual guide serves as a crucial tool to effectively apply your production plan.

Step 4: Create a planting calendar

The planting calendar plays a critical role in implementing your production plan and ensures that tasks will be carried out efficiently and in a timely manner.

1. **Finalize your production plan:** Make sure your production plan is comprehensive and aligns with your sales targets. It should include planting dates, days to maturity (DTM), harvest periods, and any specific tasks associated with each crop.

2. **Use a dynamic calendar:** With these tools, you can input planting dates, DTMs, harvest periods, and task deadlines for each crop succession. Calendar examples: Google Calendar and Excel spreadsheets with date functions.

3. **Enter crop details:** For each crop, input the following information into your calendar.
   - Planting date: The specific direct seeding or nursery seeding date.
   - DTMs: The estimated number of days between transplanting and harvest maturity.
   - Harvest period: The window during which the crop is expected to be ready for harvest.
   - Time spent in the nursery (if applicable): Note the number of days that seedlings are expected to spend in trays before going into the ground.

4. **Include specific tasks:** Add crop management tasks such as weeding, fertilizing, watering, and pest management for each crop. For instance, you may schedule tasks like "use a flex tine weeder 17 days after direct seeding" and "spray boron 21 days after transplanting."

5. **Work systematically:** Organize your calendar by week or by month, depending on different crops' growth cycles and how your farm operates. This will help visualize and plan tasks efficiently.

6. **Make sure nothing is missing:** Check that every single crop succession is included in your planting schedule, so that no crop will be overlooked, and all tasks will be scheduled appropriately throughout the season.
7. **Review and adjust the plan:** Regularly review your planting calendar to accommodate weather changes, unexpected delays, or new insights. Flexibility is key when adapting to the dynamic nature of farming.

A detailed planting calendar will allow you to establish a road map for your farm's operations and ensure that each crop receives the care it needs at the right time. This systematic approach will minimize errors, maximize productivity, and help you efficiently achieve your goals.

Step 5: Order your seeds and equipment

Once you've completed the previous crop planning and scheduling steps, the next crucial task is choosing your cultivars and ordering seeds.

1. **Select cultivars:** Choose cultivars based on their suitability for your climate, market demand, and your farm's specific conditions. Consider factors like disease resistance, yield potential, and taste.
2. **Compare suppliers:** Research and compare seed suppliers to make sure you get high-quality seeds. It's best to start the season with new seeds to guarantee higher germination rates. When using seeds from previous years, conduct germination tests to assess viability.
3. **Calculate seed quantity:** Estimate the amount of seed required for each crop, including a 30% safety margin to account for potential losses and additional seedings. This calculation should align with your production plan and planting schedule.
4. **Place orders:** Once you've chosen your cultivars and figured out how much seed you need, place orders for the entire year. Make sure the expected delivery date lines up with your planting calendar to avoid delaying any farm operations.

By carefully selecting seeds and equipment, and by planning ahead, you can optimize your farm's productivity and set yourself up for a successful growing season.

Effective Farm Management

Positive Management

Farming on a human scale is an inherently collaborative endeavor. It is a powerful example of teamwork and how it can lead to the success of a season or contribute to its downfall. According to Jean-Martin, fostering a positive attitude and cohesion within the farm community is paramount to achieving success. He often repeats the adage, "Alone you can go fast, together we can go far," and writes it on boards around the farm.

Central to the role of any grower is the ability to effectively organize, guide, and inspire the farm workforce, whether it is made up of only 2 or 3 people or a larger crew of 10 to 15. Over the years, Jean-Martin developed guidelines and principles that allowed him not only to successfully manage and mentor his teams but also to impart these invaluable lessons to his students and apprentices. At Ferme des Quatre-Temps, apprentices typically spend two seasons learning about farming: the first focuses on mastering agricultural techniques (*le geste agricole*), while the second emphasizes leadership and farm management skills.

The farm team is structured to include a core group of second-year apprentices who lead and mentor newer team members, usually six or so first-year apprentices. Jean-Martin actively collaborates with these second-year apprentices to teach them

THE MARKET GARDENER'S STEPS TO SUCCESSFUL TEAM BUILDING

1. **Include team members in discussions**
   - **Why it matters:** Involving your team in discussions fosters a sense of ownership and commitment. It allows each team member to understand their role and how they contribute to the farm's goals.
   - **How to do it:** Hold regular team meetings where you discuss farm plans and goals for the season, as well as challenges and opportunities. Encourage feedback and suggestions from team members to enhance engagement and collaboration.

2. **Give employees responsibility**
   - **Why it matters:** Delegating responsibilities based on individual skills and strengths empowers your team members. It builds trust and allows you, as the farm owner, to rely on their expertise.
   - **How to do it:** Assign clear roles and tasks to each team member according to their abilities. Provide training and guidance as needed, but also give people the autonomy to make decisions within the scope of their responsibilities.

3. **Lead by example and be present**
   - **Why it matters:** Your presence and actions set the tone for the team. During challenging times, your leadership can inspire motivation and resilience among your team members.
   - **How to do it:** Work alongside your team in the fields, especially during busy seasons or when tackling difficult tasks. Demonstrate a strong work ethic, positivity, and problem-solving skills to motivate and guide your team.

4. **Be clear and precise in instructions**
   - **Why it matters:** Clear communication helps ensure that tasks are carried out efficiently and accurately, while regular repetition will reinforce procedures and expectations.
   - **How to do it:** Clearly communicate tasks, timelines, and expectations to your team. Repeat instructions as necessary and encourage people to ask questions, to make sure they have understood. Use visual aids or written instructions to supplement verbal communication.

5. **Organize task management and crop planning**
   - **Why it matters:** Being organized will allow you to streamline workflows, keep confusion to a minimum, and maximize productivity. Good crop planning can ensure you are making efficient use of your team's time throughout the season.
   - **How to do it:** Develop comprehensive crop plans and planting schedules, as discussed earlier. Make sure that tasks are assigned and managed efficiently, considering the skills and availability of your team members. Use tools like calendars, task lists, or digital planning software to coordinate and track progress.

By applying these principles for sound team management, you can create a positive and productive work environment on your biointensive farm. This approach makes your operations more efficient, and it will also cultivate a culture of teamwork, responsibility, and continuous improvement among your team members.

leadership and help them refine their ability to guide and motivate others. His approach to team management and mentoring is encapsulated by the term "positive management," which refers to nurturing a supportive and cohesive team dynamic. This is foundational to achieving success on your farm.

Jean-Martin emphasizes several key strategies for effective farm management and team leadership. First, he stresses the importance of clarity in communication—you must clearly state expectations related to each task or chore. This clarity will limit confusion and improve satisfaction on both sides of the employer-employee relationship. As Jean-Martin sees it, frustration often arises when instructions are unclear, leading to misunderstandings and inefficiencies. Conversely, clear expectations empower employees to perform their tasks effectively and meet goals without unnecessary stress.

Second, Jean-Martin advocates for setting daily goals at the beginning of each workday. Not only does this practice help everyone align with the farm's overarching objectives but it also provides a structured framework for achieving those goals. Breaking down larger tasks into smaller, manageable steps helps maintain focus and momentum throughout the day and fosters a sense of accomplishment as each task is completed. This approach not only boosts morale but also keeps the farm operating smoothly and efficiently.

"Throughout a typical day, there are multiple tasks to manage—harvesting, topping plants, removing suckers, leaning and lowering plants, and pruning them. If you ask your team to do all these steps at once, you'll likely end up with some tasks that are done poorly or forgotten altogether. That's why it's important to give clear, specific instructions and focus on one or two tasks at a time. Make sure your team understands the instructions, then circle back later to see if everything is on track. It's better to check in frequently than to assume tasks are progressing smoothly."

A significant part of your job as a market gardener is to give instructions, and then make sure they are well understood. Once that's done, it's better to highlight good work than to complain about bad performance.

For instance, when harvesting carrots, you may ask employees to sort them into three different bunch sizes, with carrots that are roughly the same size in each bunch. Positive reinforcement here would consist of four steps.

1. Congratulate them: "Great job!"
2. Explain what they did well, thus reinforcing your instructions: "Your bunches are consistent and sorted to the right size."
3. Once again, show your appreciation for a job well done and repeat why it is important: "I'm so glad you're sticking to the instructions. Thank you! Our customers appreciate the time we spend sorting for size, especially the chefs."
4. Encourage them: "Keep up the good work!"

Additionally, Jean-Martin promotes mindfulness in task completion. Encouraging his team to focus on one task at a time with full attention enhances the quality and efficiency of their work. By practicing mindfulness, individuals can reduce errors, improve productivity, and maintain high standards of quality across all farm operations. This approach reflects Jean-Martin's belief in the power of deliberate action and mindful presence in achieving optimal results on the farm.

Jean-Martin's approach to managing his team revolves around constant reinforcement of positive work and clear communication. He believes in recognizing and appreciating good practices as they happen on the farm. When he sees someone like Anna diligently following instructions, he takes the time to acknowledge it: "Anna, when I see you bunching onions exactly as I explained, I really appreciate it. Keep up the good work."

This practice has a significant impact, boosting morale and fostering a sense of accomplishment among team members. Jean-Martin understands that correcting others' mistakes can be discouraging, so he focuses on positive reinforcement to maintain motivation and engagement. At the same time, he maintains firm standards and doesn't tolerate subpar work. He knows that consistent quality is crucial for the farm's success and customer satisfaction.

By creating a positive work environment through encouragement and clear expectations, Jean-Martin ensures that every team member understands their role in maintaining high standards. This approach not only enhances productivity but also cultivates a culture of excellence

where each task contributes effectively to the overall success of their market gardening operations.

Keep Calm and Cultivate Resilience

When things go wrong on the farm—like discovering fusarium wilt in your eggplants while dealing with weeds and spider mites—it can feel overwhelming. Jean-Martin's advice is clear: stay calm. Experienced farmers face unexpected challenges all the time. Take a moment to pause, think, and then act. Panicking won't solve anything and can affect your team's morale.

Believe in your ability to make good decisions under pressure. Rushing rarely helps; instead, approach each problem with a clear mind. Learn from setbacks, adapt your strategy, and keep moving forward. By staying calm and resilient, you will not only solve immediate problems but also build a stronger, more capable farm in the long run.

Who Should You Sell To?

One of the fundamental truths for market gardeners is this: the success of a microfarm hinges on direct sales. In other words, there is no middleman between the product and the customer. While some may argue that scaling up this model is essential to revolutionize food systems, Jean-Martin and many others in the intensive organic farming movement disagree. When it comes to agricultural business systems, they advocate for scaling back instead, to better serve local communities. This is the essence of the local farming movement.

Regardless of the broader context, farming carried out on a human scale thrives because it is deeply connected to local communities and eliminates middlemen. This leads us to the key question: which sales channels will best allow your microfarm to thrive? Throughout this book, we provide detailed insights into various sales models, exploring how acreage, staff size, and sales relationships intersect.

In his farming journey, Jean-Martin has been able to explore a wide range of models: running the initial Les Jardins de la Grelinette CSA serving 100 families, mentoring other farmers starting their own ventures, establishing iconic market stalls at Montréal's Jean-Talon market, supplying top restaurants in the city, and more. At Ferme des Quatre-Temps, various online sales channels have also been developed and tested.

Sales and Marketing

DIFFERENT SALES CHANNELS

There are different ways to sell your products directly to customers:
- CSA boxes: The concept is simple—customers pay a farmer in advance to receive a weekly delivery of freshly harvested produce.
- Farmers' markets: This is the most common but also the most competitive model. You need to make sure you have quality produce on display, your bunches look appealing, and you are bringing certain vegetables to market before your competitors (early harvests).
- Selling to restaurants: Typically, you should expect to cut your prices by 25% for restaurant sales (compared to farmers' markets). In return, restaurateurs buy large volumes and thus guarantee good sales.

Most market gardeners maintain a combination of these different income streams and sales outlets. They might sell some products wholesale, one or two kinds of vegetables, to their local food cooperative. In my opinion, combining different outlets is the best solution because you can avoid putting all your eggs in one basket. In the end, you'll still have your CSA income if one of the distribution channels shuts down, becomes less efficient, or suffers a serious blow like restaurant closures during the COVID-19 pandemic.

So which sales channel is best? Jean-Martin's experience shows that success lies in leveraging multiple channels simultaneously. Each one has its strengths and can contribute, in its own way, to the farm's sustainability and growth. Throughout this book, he shares insights into optimizing these channels for maximum impact and to better connect with customers.

"Since you're not in the wholesale business, you don't need to worry about high-volume vegetable production. You are growing small quantities of top-quality produce. The second key point to remember is that these direct sales will create a close bond between you and the customers you meet every week. You have to rely on these relationships."

In the following section Jean-Martin's insights reflect his emphasis on direct sales, community engagement, and strategic business diversification in sustainable agriculture. His approach underscores the importance of adaptability and building resilient relationships within local food networks.

Insights from Jean-Martin Fortier

CSA Boxes

"When it comes to CSA boxes, the main advantage is clear—you have a guaranteed sale before even planting a crop. This upfront payment model not only provides financial security but also allows for meticulous crop planning. When you know which sales you have committed to in advance, you can schedule plantings and harvests with precision, to run your farm more efficiently. However, with the rising popularity of farmers' markets, attracting CSA customers can be a challenge. That's why diversifying your sales channels is crucial. For a successful market garden, a loyal base of 200 to 400 CSA customers can effectively sustain the business.

Farmers' Markets

Farmers' markets offer a flexible platform where you can bring your harvest straight to consumers. Unlike CSA boxes, farmers' markets allow growers to show up with varying quantities of produce each week, which simplifies crop management. They provide a cash-driven environment and valuable networking opportunities with fellow vendors. However, they are affected by weather conditions, which impact daily sales, and vendors face intense competition. Having access to a thriving farmers' market is essential for your microfarm's success and should be a strategic consideration from the outset.

Selling to Restaurants

Working with restaurants involves building strong relationships with chefs and culinary professionals. These connections can provide the opportunity to supply produce in larger quantities and align with the growing farm-to-table movement. Challenges with this clientele include the occasional late payment and the unpredictability of demand, due to changing menus or restaurant closures. Despite these risks, partnering with restaurants can be rewarding, offering a steady outlet for your high-quality, locally grown produce.

Set Your Prices Right

When setting prices for your vegetables, focus on their market value rather than perceived value. Aim to consistently price your produce at the higher end of the market range. To determine your rates, I recommend looking at prices posted by other vendors at farmers' markets and checking out the organic section of your local grocery stores. This comparison will give you a clear understanding of what fresh, local organic vegetables typically sell for. Given the high quality of your produce, set your rates in the upper range of these benchmarks.

You'll know you've set your prices right when a handful of people complain about them. This feedback means that you're likely capturing the true value of your produce. Conversely, if no one is questioning your prices, you may not be charging enough.

As a farmer, your livelihood depends on these sales, so it's crucial not to undervalue your products. Your goal is not to make organic vegetables affordable for everyone—it is to make sure you receive appropriate compensation for your hard work and dedication. Instead of setting prices based on personal preferences, focus on aligning with standard market rates for maximum profitability.

Reflecting on my own experience, I once sold carrots at Can$2.50 per bunch while my competitors sold them for Can$3.50. Over one season, that small difference cost me Can$2,500 (US$1,800) in potential revenue. Setting prices slightly higher can significantly increase your income without requiring additional effort in production, packing, or harvesting.

Finally, maintain consistent prices throughout the season to discourage negotiation and set customer expectations.

Signage and Presentation

Use clear signage for each product, ideally with chalkboards attached to rotating label holders, which is a professional and reusable solution.

Consider displaying your vegetables in wooden crates for a visually appealing setup that enhances their marketability.

High-quality produce also requires proper packaging. For instance, salad packaging should be fluffed up and filled with air to maintain freshness. But the most important strategy is to stack your produce as high as possible to create a sense of abundance, a striking visual that will draw customers to your booth. A key trade secret for successful farmers' markets is to build towering stacks of vegetables piled up to eye level—they will have a powerful impact. Keep rotating your produce, regularly placing fresh items on top, so that things look crisp. On warmer days, you can use a spray bottle to make everything look shiny and keep the vegetables looking vibrant.

Customer Service

When it comes to customer service, every detail matters. Your sales team's appearance is critical. Make sure they've washed their hands, are well-dressed, rested, and wearing their aprons—it adds a professional touch! Hiring people who smile is key. A top tip for building customer loyalty is to remember their names. Greeting returning customers by name helps forge strong relationships. It shows you value them beyond just a transaction.

Keep chairs away from your stall; serving customers while standing will keep your team engaged and accessible. If you need to bag produce, set up your station so that it faces the customers. Stay present, attentive, and welcoming.

Resist the temptation to use your phone, whether it is for personal calls or orders. It sends the wrong message and distracts from customer interactions.

And crucially, never argue with a dissatisfied customer. If there's an issue, calmly offer a refund or replacement. In my experience, arguing rarely resolves problems and can harm your reputation.

In short: no phones, no chairs—"yes" to aprons, remembering names, politeness, and always smiling. These principles are the foundations of excellent customer service, proven to build customer loyalty and satisfaction.

Upgrade Your Payment System

Upgrading your payment system is a crucial step in optimizing your farm's revenue streams, especially at farmers' markets. Jean-Martin highly recommends transitioning to a card payment system like Square.* Customers who pay by card typically spend more than they would with cash, making the small percentage fee per transaction worthwhile in the long run.

With payment software like Square, you can streamline your checkout process. It allows you to view inventory, generate invoices using

* https://squareup.com/us/en

just a scale and tablet, and manage transactions efficiently. You'll gain comprehensive sales tracking capabilities, so you can create detailed reports and run analyses by location or product. With this data, you can more accurately predict sales and plan your harvests effectively, based on market trends.

Consider implementing a loyalty card program to enhance customer retention. Offer discounts or gifts after a set number of purchases to encourage repeat business and secure sales in advance. Additionally, use incentives on your website to presell a portion of your harvest through prepaid credits that customers can redeem directly at your farm.

To expand your customer base and foster loyalty, launch a promotional campaign for your loyalty cards. Offering small discounts to loyal customers not only encourages return visits but also stabilizes sales over time.

Profiles: A Look at 8 Successful Microfarms

Les Jardins de la Valette

Sylvain Couderc

FRANCE

Sylvain Couderc was born in 1983 and grew up in Aveyron, France, in a family that has agricultural roots (his grandparents were farmers). He always wanted to work the land and grow plants, with a job that would bring in a decent salary while allowing him to develop practices that better protect the environment and biodiversity. When he was younger, Sylvain would spend hours in the family garden, drawing inspiration from permaculture. But he went in another direction with his studies, instead choosing to get a French BTS, or technical diploma, in electrotechnology. His return to the land began when he enrolled in a vocational program leading to a nurseryman diploma, and was later cemented when he spent a year learning the trade on agroecological farms in Australia and New Zealand. "I wanted to explore the most cutting-edge systems in these countries, but I couldn't find a viable model for permaculture. And 12-acre organic farms running on tractors didn't satisfy me either. But that's where I first came across people using Eliot Coleman's methods. So I decided to continue my agricultural training with a BTS in horticulture," he explains.

As a part of his studies, in 2010, he completed a one-month internship with Jean-Martin Fortier and Maude-Hélène Desroches. "When I walked around Les Jardins de la Grelinette, it was a revelation. Here, in real life, I was seeing everything I had read about in Coleman's work. The system appealed to me right away: the layout, the size of the farm, the quality of the vegetables…an organized garden, square, aesthetic, not too big. I wanted to get home as quick as possible, to do the same thing in France." He cut his BTS studies short, and with the help of the Aveyron Chamber of Agriculture, Silvain set up a diversified market garden on his family's land (1 acre of farmed land, over 3.7 acres). "I was lucky to have access to land, which was a key factor that made the start-up easier. This wasn't an ideal spot for market gardening, with clay and limestone soil, cold weather in the spring, and very limited water resources. But I didn't have to rent land or housing. I don't know if I would have managed to get established without that boost. I was able to start with a €20,000 (US$21,500) grant for young farmers and a €40,000 (US$43,500) loan.

FARM STATS:
LES JARDINS DE LA VALETTE

SIZE OF THE FARM

7.5-acre farm
1.2 acres usable farm area
0.3 acres high tunnels
Divided into 120 beds, each 160 ft^2

SALES

2022: €75,000 (US$81,000), taxes incl.
2021: €62,000 (US$67,000), taxes incl.
2020: €55,000 (US$60,000), taxes incl.

YEARS OF OPERATION

Since 2012

MARKETING & DISTRIBUTION

Direct sales at 2 markets (90% of revenue)
On-farm sales (10% of revenue)

VARIETAL DIVERSITY

45 vegetable species and 105 varieties

NUMBER OF EMPLOYEES & HOURS WORKED

2 FTE (full-time equivalents) spread over 1 year:
• 1 employee at 32 hours per week, March to September
• 1 employee at 16 hours per week, October to January
• 1 intern at 32 hours per week, March to September

INVESTMENTS (EXCL. LAND & REAL ESTATE)

Total invested since the beginning: €120,000 (US$130,500)
€60,000 (US$65,000) at the start
€40,000 (US$43,500) after 5 years (move to new farm)
€10,000 (US$11,000) after 7 years (new greenhouse)
Investments made over the past 2 years:
€10,000 (US$11,000) to drill a well
€10,000 (US$11,000) for equipment (new irrigation,
    new cold room, tools)

That allowed me to pay for tools, irrigation, a greenhouse, a tiller, a flail mower, a harrow, a delivery truck, and a cold room. This was in 2012, when Jean-Martin's method was still relatively unknown, and he had not yet published a book. I kind of 'free-styled' the start-up, but I felt quite confident."

In the first few months, Sylvain relied on invaluable help from his father and his partner, Nancy, who was still a student at the time. "That support was crucial because getting started alone is hard. I religiously applied what I had seen and learned with Maude-Hélène and Jean-Martin. In the fourth year, I hired someone for a few hours a week, from March to December, and increased those hours every year. In 2017, I hit my €40,000 (US$43,500) sales target, which meant I could also earn a minimum wage. I just divided the surface area and number of workers by two: at La Grelinette, 3 full-time employees work on 2 acres of land, and my operation, at Les Jardins de la Valette, is half of that."

Moving the farm

In 2017, Silvain and Nancy, who had become a teacher, began to consider acquiring new land and, especially, a place to live. "We each had a salary and could take out a loan. We were looking for a farm within a 30-mile radius of our current location, so we could stay close to our existing distribution channels. I got my first loan, €40,000 (US$43,500) over seven years, for the market garden (greenhouses, irrigation, retention pond). The second loan I took on with Nancy, to purchase land and buildings (€230,000 [US$250,000]). This new iteration of the farm showed me that the Fortier system is easy to replicate; it took me only three months to set up beds and build greenhouses. As a result, I was able to maintain a continuous farm output, and by 2018, I was harvesting my first vegetables. On the second farm, you don't make the same mistakes you made on the first one—you do it all over again, but better. This might not be the best soil for market gardening, but it's still a great Aveyron soil, sitting 1,500 feet above

sea level. It's an acidic sandy loam that's easy to cultivate. More importantly, we now have water autonomy, thanks to a spring-fed well and a 100-foot well that we had dug to fill the pond." Sylvain even managed to increase his soil organic matter content, from 1.5% to 4.3% between 2017 and 2021, after he consistently added green waste compost to his beds. This organic matter mineralized the soil and made it easier to cultivate, which in turn improved weed management and increased yields. With better soil quality and more water, Sylvain's sales shot up from €44,000 (US$48,000) in 2018 to roughly €75,000 (US$81,000) in 2022.

A marketing approach that relies on two weekly markets

Les Jardins de la Valette is divided into 7 plots that are in a 7-year rotation, with each plot containing 12 beds. The 6 greenhouses are in a 6-year rotation, each containing 7 beds. "Here, we don't heat the greenhouses, our climate is more forgiving in that respect, and we start going to market one month earlier, in April. We plan everything so that our last vegetables will be sold at the end of December. Then, we empty our greenhouses in January, clean them, and immediately fill them with plants ordered from organic nurseries: mâche (corn salad), lettuce, cabbage, spinach, beets, arugula, turnips, radishes, Swiss chard, fennel, parsley, pac choi cabbage, kohlrabi. After a one-month break in February, I start again in early March with seeding and planting."

With the 20 or so crops that reach maturity at the end of March, Sylvain can stock a beautiful farm stall in early April. He brings this produce to his flagship market in Limognes-en-Quercy, in Lot, and to a "very promising" market that he just recently joined in Villefranche, in Aveyron. "Until recently, I had been selling CSA boxes, but I decided to give that up because it takes a lot of effort for mixed results, especially with a rural clientele. I prefer to focus on this second market, because I like the atmosphere around the stalls and the relationships I build with customers and colleagues."

KEY DATES

2008
Trip to Australia and
New Zealand, discovering
Eliot Coleman's methods

2010
One-month internship
at La Grelinette

2012
Starting up on 1 acre
of family land in Aveyron

2017
First time Sylvain
earns minimum wage
Setting up in a second
location, after purchasing
farmland and a home

2022
Joining a second market,
€75,000 (US$81,000)
in sales

THE BIOINTENSIVE SYSTEM, ACCORDING TO SYLVAIN

This model changed my life. It is a mindset based on organization and planning, and this approach to overall farm management makes it super efficient. The ultra-rigid and structured model speaks to me, it works for me. It allows me to maintain my family life, earn a decent salary for 40 hours of work per week, and enjoy two months of vacation per year, one month in February and four weeks spread throughout the year. All this without a major cash contribution and few loans. If I hadn't gone to Maude-Hélène and Jean-Martin's farm, I probably would not have become a market gardener. I was familiar with permaculture and grew crops according to those principles, but I didn't see how I could earn a living doing that. With the biointensive system, I realized that it was actually possible. In future, I want to be more efficient every year, stay the same size, and continue doing what I enjoy.

SYLVAIN'S ADVICE

For many people, it can be hard to stick to a model because everyone wants to create their own farm in their own way, even though it would be better to follow certain principles. It took me a decade to accept this and to apply the model a little more every year. There's always this temptation to run around, left and right; it's a mental game, you have to stay focused. Don't they say that "the greatest liberty is born of the greatest rigor"?

It's helpful to have multiple sales outlets from the start, even when your output is still quite small. By doing this, you'll be able to fine-tune your distribution channels and focus on the ones that are most efficient, most profitable, and easiest to organize. The diversified market gardening model really is best suited to direct sales. Things are more complicated with stores that push down prices to improve their margins.

Every season, I make a crop plan using open-source software. With this plan, I know what needs to be planted and where, every week, without having to second-guess myself. It tells me the crop successions I have to plant, which saves me time and ensures that I'll always have enough produce for my customers.

Abricot
€
1.9d

Une Ferme du Perche

Tom Rial

FRANCE

The story of Tom and his farm began in 2000. This is when his father, Jean-François Rial, a passionate entrepreneur and founder of Voyageurs du Monde, bought a second home in Le Perche, Normandy, not far from Paris. He hoped this would make his children more aware of nature and give them the taste for the countryside. "From a young age, I would come here with my family every weekend, and that's probably where my desire to work outdoors comes from," explains the young Parisian. Despite this inclination, Tom went on to study marketing in New York City, where he quickly developed an interest in agroecology. "My introduction to agriculture started when I discovered the Bec-Hellouin farm, in Normandy, and read *Permaculture* by Perrine and Charles Hervé-Gruyer." Tom was hooked on market gardening, so he read and watched everything he could on the topic.

While Tom was working his first job in Montréal, his father came to visit, and they met Jean-Martin at a dinner. "We hit it off right away. The next day, my father toured Ferme des Quatre-Temps, and that's when he realized that it was actually possible to run a profitable vegetable microfarm. By that point, I was tired of living in the city and was feeling disconnected from my work in advertising, a job that had no impact on society. I wanted to create my own farm, with support from Jean-François, who became my business partner." Tom flew back to France and completed a two-month training program at Ferme Sainte-Marthe, in Sologne, a pioneering organic farm. "The course was useful, but I still needed to learn more before starting my own market garden. I highly recommend that people learn the ropes, get an agricultural diploma like a professional farm management certificate (BPREA), and work several seasons. This is something I was lacking throughout the start-up years. But I also think it takes some amount of ignorance and a dash of madness to even dare to become an entrepreneur and market gardener!"

FARM STATS:
UNE FERME DU PERCHE

SIZE OF THE FARM

12-acre farm
2.5 acres unheated multi-span greenhouses
0.5 acres caterpillar tunnels

SALES

2021: €250,000 (US$270,000), taxes incl.

YEARS OF OPERATION

Since 2019 (preparing the land)
First season in 2020

MARKETING & DISTRIBUTION

Direct sales in 4 markets (45 % of revenue):
• Bellême market (Orne), Thursdays
• Président Wilson market (Paris), Wednesdays and Saturdays
• Mortagne-au-Perche market (Orne), Saturdays
• On-farm sales "les mercredis fermiers" (farmer Wednesdays): annual summer market for local growers and artisans

Sales to businesses (35% of revenue)
• 20 restaurants (30% of revenue)
• 3 grocery stores (4% of revenue)
• 1 bar (1% of revenue)

CSA sales — 100 boxes (20% of revenue)
• Paris (60% of CSA revenue)
• Local (40% of CSA revenue)

VARIETAL DIVERSITY

50 vegetable species and 150 varieties

NUMBER OF EMPLOYEES & HOURS WORKED

5.5 FTE spread over 1 year (5 or 6 employees in season / 3 in the winter)
Payroll: €130,000 (US$140,000)

INITIAL INVESTMENTS (EXCL. LAND & REAL ESTATE)

€350,000 (US$380,000) without loans or agricultural grants (including greenhouses, irrigation, tools, wash station, and cold rooms)

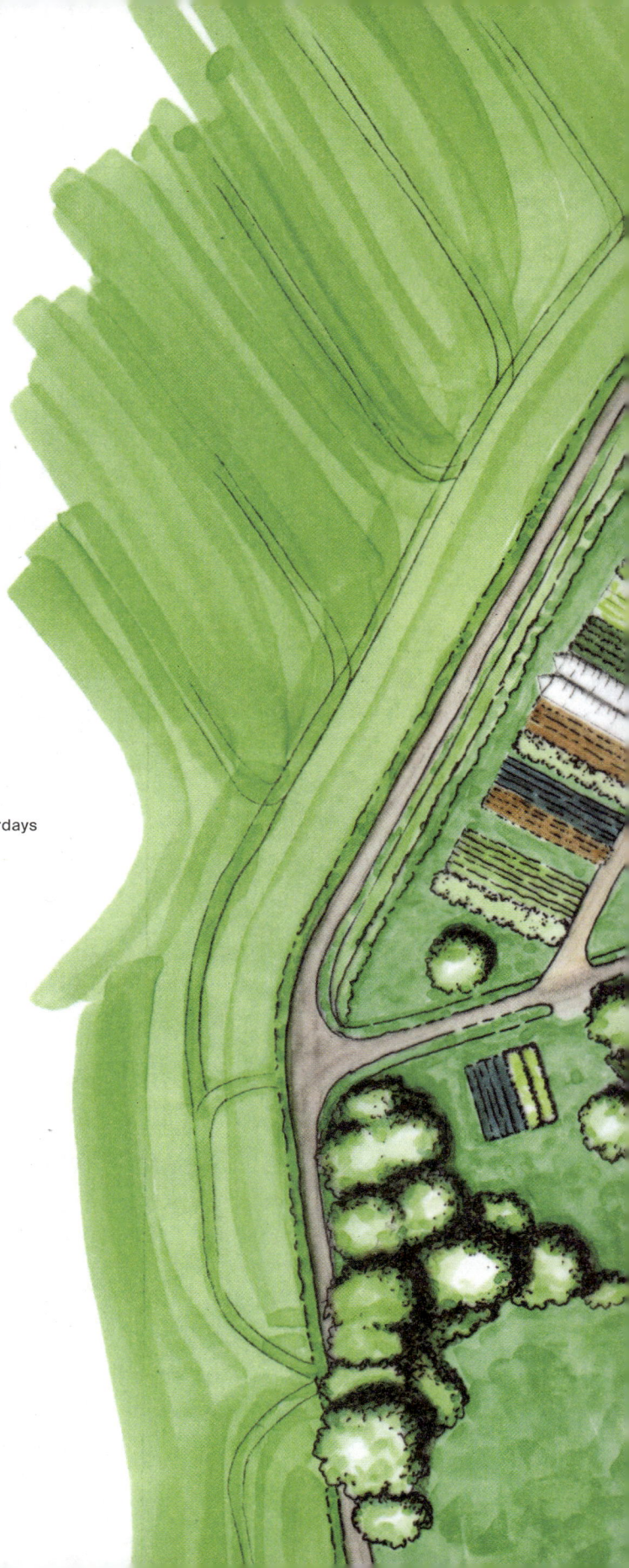

0   25   50   75   100 m

A family- and community-run adventure

In 2018, Jean-François bought 12 acres of land from a farmer who was about to retire. "It used to be a pasture for dairy cows, and it was right next to our second home. The lot came with advantages, like highly fertile clay-loam soil with a high organic matter content (5%), and disadvantages, such as hilly terrain that required expensive earthworks, not to mention the time lost to travelling between the two properties."

When Tom and Jean-François first started the farm, Fermes d'Avenir, a French organization dedicated to supporting agroecological farming, provided guidance, especially when it came to recruiting Louise, their farm manager. Jean-Martin also supported the farm, which he visited for the first time in November 2018. He told them, "If I find earthworms, I'll take on this project with you." Armed with a spade, they dug up one scoop of soil and found dozens of them. While this was initially a small permaculture farm project, it quickly grew into something more ambitious: Jean-Martin went on to officially support the project and, to design this market garden, he reached out to Québécois landscape architect Alexandre Guertin, a specialist in ecological polyculture farms. "In 2019, we created the first field blocks, invested in earthworks, installed irrigation systems, and recruited a second employee, Gauthier. Together, he and Louise flew to Quebec to learn from Jean-Martin. I also visited a number of farms with him. When we got back, we kicked off a mini season with vegetable boxes for employees at Voyageurs du Monde, which brought in €5,000 (US$5,500). Our first full season was in 2020, and that's when we made significant investments to set up employee housing and a wash station, and to build our multi-span greenhouse. To create a vibrant and fertile space, we planted more than 5,000 trees, set up a nursery, dug a pond, built 10 tunnels to grow crops under shelter, and amended our beds every year with compost, green manures, and nitrogen fertilizers like castor meal and chicken manure."

The Ferme du Perche adventure may have started out with only two people, but it soon became a collective project. In just 3 years, the farm had a team of five employees between the ages of 25 and 42, all committed to cultivating these 2.5 acres on which they grew—depending on the season—heirloom tomatoes, eggplants, carrots, radishes, cabbage, spinach, salad greens, onions, cilantro, ginger, etc. "Louise, Gauthier, Masami, Guillaume, Mathilde, Antonin, Julie, Camille, Vanessa, Baptiste, Barry, and Selma have all spent at least one season on our farm." Every day, they work hard to put all that they have learned into practice. For some, these skills and knowledge were acquired during their farm stay in Quebec. "We see our farm as a way of life—we all sleep on-site, we share our work and our meals. Most of us hail from major metropolises across France and came here looking for a purpose, a project that would connect us with nature. We were driven by this need to move away from a daily reality that had become unbearable to find something on a more human scale. Almost all of us have a Bac+5 [graduate degree] and could earn a salary commensurate with our level of education. But we pour everything we have into this profession that answers our common search for meaning. Here, we feel useful. Farms can help rebuild that bond with nature, can become a space for everyone, so we can show people what we do, and also organize events. This farm is more than just a place where we grow food, in our own little corner. It's also a place for sharing."

TOM'S ADVICE

To learn about crop management, I recommend the Market Gardener Masterclass as well as the three volumes of *Produire des légumes biologiques*, a technical handbook published by the French Research Institute of Organic Agriculture (ITAB).

From the very beginning, you should design the farm as if it were in full swing, but you also need to start as small as possible. We prepared all our field blocks at the same time, and that was way too much; we quickly became overwhelmed. You should begin with one well-managed garden, which is better for morale and for your work. Then, you can slowly progress once you're in control of the initial bed space and expand as you go. I suggest prepping 10 beds, tarping most of them, and then opening them up gradually.

A significant rise in sales

Tom spends half of his time on management, sales, and communications and the other half on field work. He sees market gardening as "a material handling job: you carry materials into the ground and you pull others out of it... We're nature's movers!" During the first COVID lockdown in the spring of 2020, he teamed up with other local growers to offer *paniers perchés*, farm boxes delivered to consumers in their region. "It was madness, we were assembling 100 boxes a week," he recalls with enthusiasm. "We then recruited Mathilde and Guillaume for more support, and started our first market, in Mortagne-au-Perche, in the summer of 2020. We made €95,000 (US$103,000) in sales that season. In 2021, Louise and Gauthier left to start their own farm, and Julie, Antonin, and Camille joined us. We launched our CSA program, started going to two new markets in Paris and Bellême, and brought in €52,000 (US$57,000) in sales. We also started running a market on the farm,

called '*les mercredis fermiers*' [farmer Wednesdays]. Since then, we have also been selling our produce to prestigious restaurants like Plaza Athénée, by Jean Imbert, and Halle aux Grains, by Michel and Sébastien Bras."

In 2022, Baptiste and Vanessa Saulnier came onboard. "This was a game changer because they already knew the market gardening process inside out. Baptiste had learned the trade at Ferme des Quatre-Temps, and with Eliot Coleman. Both he and Vanessa arrived with a wealth of experience as farm managers and had been in charge of setting up a biointensive market garden at Château de Chambord. This year, we plan to bring in €250,000 (US$270,000) in revenue with €130,000 (US$140,000) for payroll, €60,000 (US$65,000) for fixed costs, and €30,000 (US$32,500) for investments."

Ferme du Perche was financed with our own funds, without any loans or start-up grants. After three seasons and four years of operation, the farm turned a profit.

KEY DATES

2000
The Rial family buys a second home in Réveillon, Perche

2016–2018
Tom goes to school in New York, gets his first job in Montréal

2018
The Rials meet Jean-Martin, visit his farm, and purchase Ferme du Perche, in Réveillon

2019
Farm start-up, mini season with some boxes, €5,000 (US$5,500) revenue

2020
First full season, multi-span greenhouse is built, first markets, €95,000 (US$103,000) revenue

2021
€152,000 (US$165,000) revenue

2022
Farm becomes profitable with €250,000 (US$270,000) in revenue, including roughly €30,000 (US$32,500) of annual depreciation

THE BIOINTENSIVE SYSTEM, ACCORDING TO TOM

This is a fantastic way to cultivate organized and rigorous practices. With the biointensive system, you can farm a smaller area, and thus be much better organized. We work outside, to the sound of birds and bugs. But, at the same time, our operations are mainly guided by spreadsheets, with processes that we follow to a T. They cover everything, from harvests and eggplant pruning to general planning, meals, newcomer training, etc.

With this approach, we can be economically efficient and maintain a good quality of life, working only 7 hours a day (8:30 a.m. to 4:30 p.m., with a lunch break), or 35 hours per week (annualized). We work longer hours in the summer, and are on call on weekends. The system also allows us to take five weeks of vacation per year: one in the summer, one in the spring, and three in the winter, plus one month off for overtime

accumulated in the summer. With our farming system, we run a profitable operation and can pay market gardeners a decent salary. When our employees leave us, they have enough experience to start their own farm. Our mission is to promote openness, sharing this approach with as many as possible. As more people embark on this journey, our ecosystems will improve.

Le Potager des Ducs

Mathieu Lotz

FRANCE

Le Potager des Ducs is an urban agriculture project that runs on a hybrid model, growing vegetables and microgreens—a nutrient-dense and flavorful product. Alsatian Mathieu Lotz founded the project in February 2018, when he was given the opportunity to set up shop in the municipal greenhouses of Dijon, France. He offered his produce to restaurateurs and organic grocery stores in the city.

An engineer by training, Mathieu had worked in the production logistics industry for eight years when he took this abrupt turn. "For a few years, I worked in my field of study and got to lead great projects. But eventually I felt compelled to shift my professional focus to align it with my values, bring more meaning to my daily life, and get out of the factories to see sunlight again! So I visited a few farms in France to get a better understanding of this new industry, even though I didn't have an agricultural background at all." At this point, Mathieu decided he needed professional retraining, and started to pursue a BPREA, or agricultural business management diploma.

Starting up on city land

Matthew knew very little about how the agricultural industry worked in his new hometown, so he began to learn about local growing practices and to meet the market gardeners who would be his future colleagues. "I was in school, and used this opportunity to get internships both on farms that were in the start-up phase and on big well-established operations. I wanted to take a year to explore several models," he explains. In 2017, with his BPREA diploma in hand, Mathieu worked with the Côte-d'Or Chamber of Agriculture to finalize his small-scale farm plan, and gained experience while employed, for a full season, on a market garden near Dijon. "I was able to see the different vegetables he was growing on 17 acres, the way he organized his day-to-day work, and his marketing and sales processes. He is a solid market gardener who taught me a lot about management, technical decisions, and how to be economically viable. On his farm, I was also able to practice working with predominantly clay-limestone soil, which was similar to the soil where I hoped to set up my farm."

FARM STATS: POTAGER DES DUCS

SIZE OF THE FARM

- 0.75 acres of farmed land, split into 0.3 acres under shelter (32 beds, each 100 × 2.6 ft)
- 0.1 acres (5,500 ft² of high tunnels for vegetable seedlings, microgreens, and mushrooms that also include storage space, washing and packaging stations, and a workshop. This area is kept above freezing temperatures.

SALES (INCL. TAXES)

2021: €99,000 (US$107,000) includes €62,000 (US$67,000) in vegetable sales
2020: €78,000 (US$85,000) includes €46,000 (US$50,500) in vegetable sales
2019: €70,000 (US$76,000) includes €30,000 (US$32,500) in vegetable sales
2018: €23,000 (US$25,000) includes €5,000 (US$5,500) in vegetable sales

YEARS OF OPERATION

Since 2018

MARKETING & SALES

Direct sales to grocery stores and organic shops (50%) and restaurants (50%)

VARIETAL DIVERSITY & BEST CROPS

Roughly 30 species and 120 varieties
Most successful vegetables:

- 8 heirloom tomato varieties (about half are grafted at the farm)
- 3 cherry tomato varieties (3 colors)
- Salad mix made with different varieties of salanova
- 4 zucchini varieties (long and round, green and yellow)
- Bunched beets: red, gold, chioggia, and white (different sizes depending on the needs of restaurants and grocery stores)
- Purple and graffiti eggplant

NUMBER OF EMPLOYEES & HOURS WORKED

2.1 FTE spread over the year:

- farm manager: 45 hours/week over 45 weeks, approximately 2,000 hours
- 1 permanent employee: 1,600 hours + seasonal: 300 hours
- Interns: 400 hours

INVESTMENTS (EXCL. LAND & REAL ESTATE)

2018–2019: about €21,000 (US$23,000), including €18,000 (US$19,500) from the DJA (fund to support young farmers)
2020–2021: €36,000 (US$39,000) over the years (personal funds)

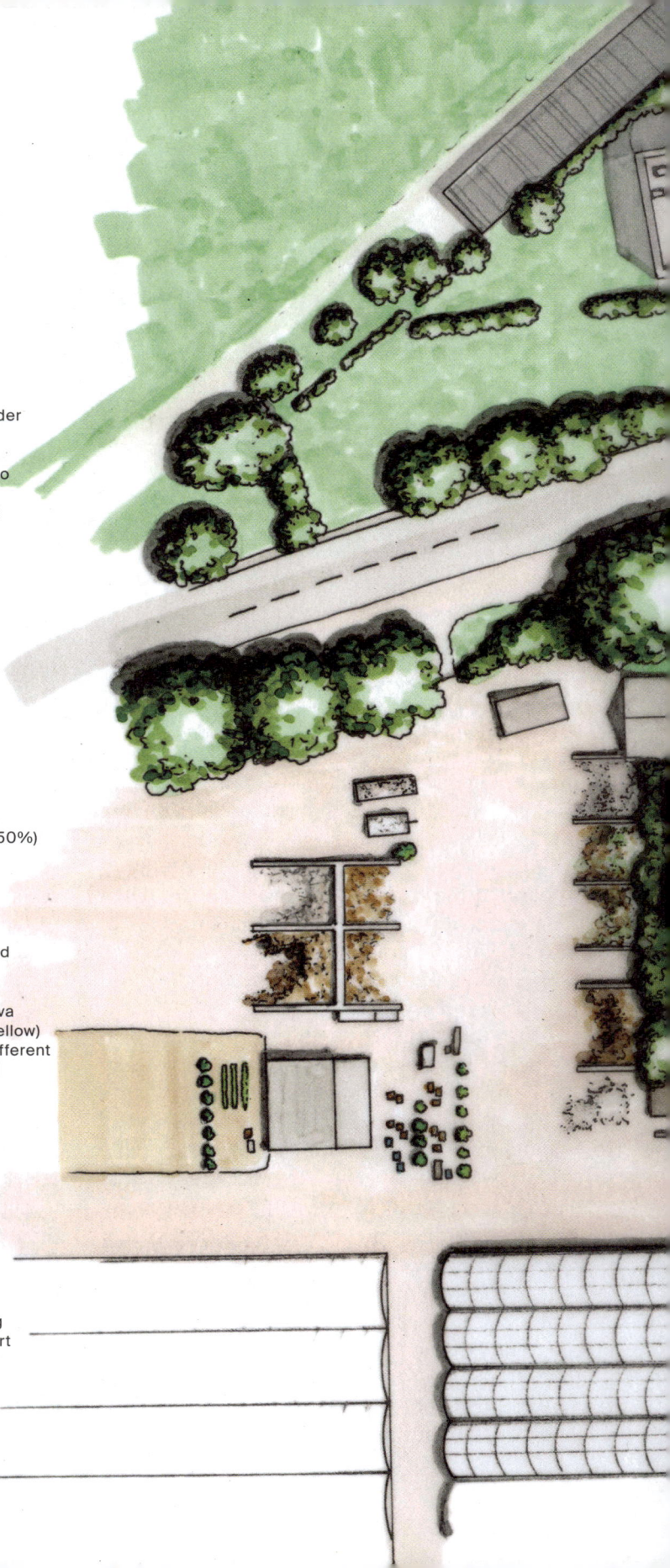

N

0   10   20   30   40 m

At the same time, Mathieu was looking for land and visiting urban farms in other cities, for inspiration. He became particularly interested in projects focused on innovation. "I wanted to align myself with the values of new small-scale farmers like Jean-Martin and Curtis Stone. At one point, things just clicked for me, especially when I visited Cycle Farm, a small operation south of Brussels, and realized I needed to find land in an urban or peri-urban area," says Mathieu. So he took to the sky, with help from Google Earth, to find little green areas in Dijon. He explains, "That led me to where I am today, just over a mile from the city center, near vineyards in the Bourroches neighborhood." He is leasing a plot on a bigger piece of land that is dedicated to horticultural operations for local parks and gardens. "I've slotted myself in there, on unused land, where I farm 0.75 acres in the field. I also have 5,500 square feet of space in a high tunnel with a ground cover. Today, I use that space as a packing and washing station, storage area, and nursery. It's my farm's nerve center. We keep the space above freezing in the winter, which means I can grow 100% of my seedlings on the farm." Mathieu followed the classic start-up journey and got a grant for young farmers through the French DJA program, though not without some challenges, "like justifying a full-time salary and proving this was more than just a hobby farm, despite the small footprint." He ended up ticking all the boxes and received €18,000 (US$19,500) in 2018. But his land and building lease is subject to a precarious five-year occupancy agreement, and he admits that's "a stressor, it means I haven't really secured land." Still, he feels optimistic about the future, because "Potager des Ducs gets support from the City, and they regularly publish statements about this ultra-local organic farm project."

Careful optimization

In the first year, Mathieu decided to grow crops on only a portion of the available land, "to avoid becoming overwhelmed by field work." This gave him time to focus on finding outlets for his first harvests. "We started with 0.2 acres, and I built our first 6,000 ft^2 multi-span tunnel (2 tunnels with a shared wall) in 2019, followed by a second one in 2020." With this significant amount of bed space under shelter, he generated better yields than he would in unsheltered field blocks. Mathieu gradually worked his soil with a walk-behind tractor, a power harrow, and a broadfork, "to aerate and loosen this relatively heavy clay-limestone soil and create standardized beds, each 100 feet long and 30 inches wide." Once or twice a year, he spreads compost over the beds, "to add organic matter and improve the soil structure."

To optimize such a small growing surface, Mathieu has chosen to limit aisle space in the tunnels when growing winter vegetables like Chinese cabbage and salads greens. "We also alternate crops with short and long growing cycles. Vegetables with a quick turnaround include pac choi, which is in the field for 30 days, salads greens, and sucrine romaine. We bring them to market fast and without much storage time. Slower crops, like zucchini and tomatoes, stay in the field for longer and are harvested multiple times. Once that's done, we cover the ground for winter (December to February) with black silage tarps. This helps us better manage weeds in the spring and warms the ground before we dig into a new season.

MATHIEU'S ADVICE

It can be useful to take out a small loan so that you have working capital. This will give you some flexibility with cash flow in the start-up phase and alleviate pressure when the time comes to harvest your first crop.

We can never say it enough, but start SMALL! In the beginning, you have to be able to grow produce, gain confidence in your work, learn to lose some crops and, especially, manage to sell what you've harvested.

We're quite isolated on our farms, so I find that delivering your own produce to customers is rewarding. It feels good to go to the restaurateurs and grocers, to tour their kitchens, and talk with the chefs.

Marketing and sales focused on salad greens, tomatoes, and microgreens

Today, Potager des Ducs grows a significant amount of produce: "We manage to harvest 12 to 13 tons of vegetables every year," says Mathieu. "We have a range of key crops that help bump up our revenue, and these are primarily summer vegetables like heirloom tomatoes and cherry tomatoes (our heavy hitters, bringing in 30% to 40% of our revenue) as well as zucchini, eggplant, and leafy greens, such as our mix of 4 or 5 salad varieties. We also grow small yellow, red, and white beets for restaurants."

Before getting to this point, Mathieu started out with a focus on microgreens, "to build capital more quickly. With this approach, I was able to knock on far more doors, and later add field crops to my offering." Growing microgreens was an idea inspired by many successful farms like those run by his North American mentors, Curtis Stone and Ben Hartman. "In early 2016, I started watching their videos on YouTube. I found it interesting that they would grow these early crops under shelter to complement their vegetable production. In urban areas, microgreens provide added benefits: they have a small footprint, can be harvested year-round, and provide access to a particular clientele. It's an appealing product that can help you get a foot in the door with restaurants, without any major initial investments. I'm fortunate to have access to greenhouses, in which I was able to quickly start growing microgreens. That added to my revenue, meaning there was less pressure when it came to vegetable harvests."

Mathieu is not managing all these operations alone. In 2020, Sébastien joined him, first as a full-time employee and soon after as an associate, under the GAEC program, a French system that supports collective farming initiatives. "The idea was to share the responsibilities on the farm and to generate a better hourly income." As a result, Mathieu had more time to focus on microgreens production, marketing, administrative tasks, and deliveries, while still "available to help in the garden."

With the pandemic and restaurant closures, Mathieu and Sébastien took a new tack, trying to establish sales channels through grocery stores. "That change has served us well, as our current revenue is split 50/50 between restaurants and grocery stores. We collaborate with people who share our values: Au Gramme Près, Locavore des Bourroches, Papilles, and L'Épicerie Paysanne."

To stabilize their operations, which are running smoothly now, and to avoid relying only on summer vegetables, which can suffer the vagaries of pests or weather, Mathieu has already run tests and is planning to start growing mushrooms: oyster, shiitake, cremini, and exotic mushrooms. "This crop is well suited to an urban environment and fits into our off-season, so we'll have it set up in the greenhouse from October to April. Investing in a new workshop will also make the farm more resilient."

Potager des Ducs thus continues to diversify, with a guaranteed, reliable distribution network of restaurateurs who are already interested in their upcoming mushroom offering.

KEY DATES

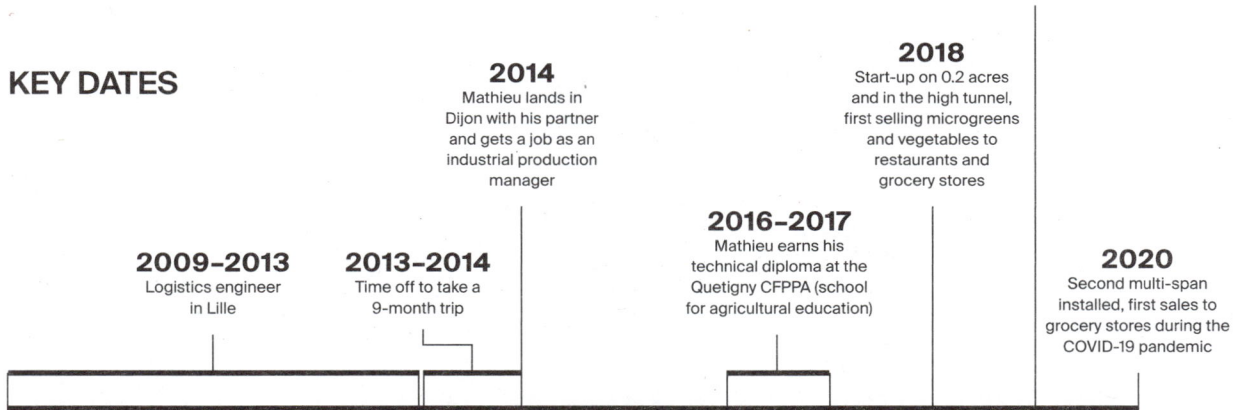

2009–2013
Logistics engineer
in Lille

2013–2014
Time off to take a
9-month trip

2014
Mathieu lands in
Dijon with his partner
and gets a job as an
industrial production
manager

2016–2017
Mathieu earns his
technical diploma at the
Quetigny CFPPA (school
for agricultural education)

2018
Start-up on 0.2 acres
and in the high tunnel,
first selling microgreens
and vegetables to
restaurants and
grocery stores

2019
First multi-span tunnel
is built, field blocks are
prepared in two thirds
of the field

2020
Second multi-span
installed, first sales to
grocery stores during the
COVID-19 pandemic

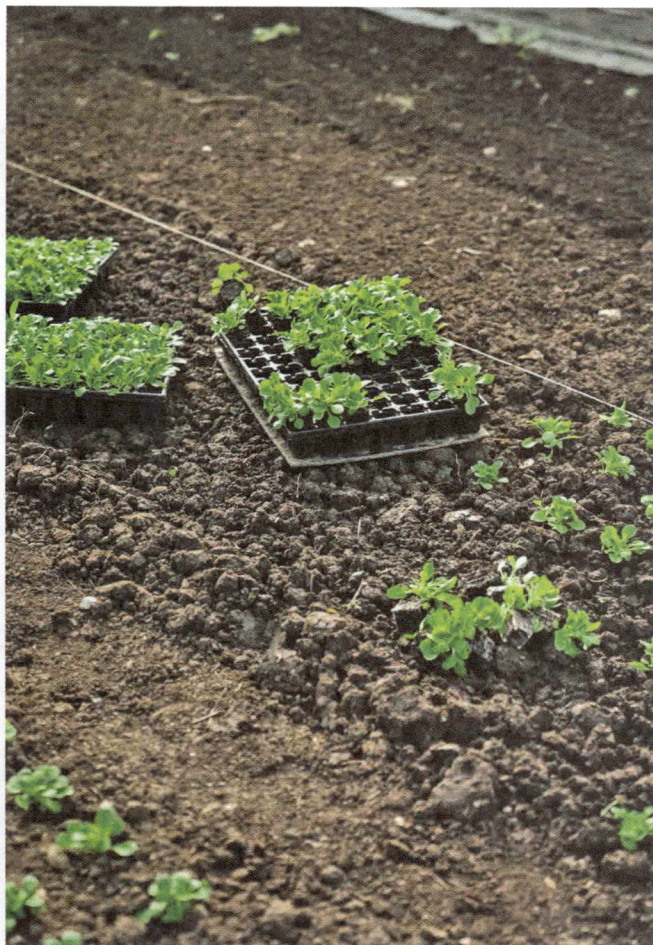

THE BIOINTENSIVE SYSTEM, ACCORDING TO MATHIEU

The biointensive approach is well suited to urban and peri-urban agriculture, where access to land is quite limited. With a minimal surface area, growers can generate high yields and good sales, thanks in part to the proximity of the farm's clientele.

Les Jardins de la Banquise

Suzanne Esteve and Antoine Régeard

FRANCE

The name Les Jardins de la Banquise, which roughly translates to "Ice Floe Farm," is like a mantra, a talisman, an alliance with a symbol of climate change, to better protect against the relentless reality of rising waters. It especially speaks to the humor of the farm's founders, Suzanne Esteve and Antoine Régeard, who have been together for eight years and are ready to take on the wild challenge that is growing diversified vegetables from open-pollinated and heirloom seeds in a flood zone of the Gironde estuary in southwestern France. They met in 2014, in Israel, when Suzanne was studying international migration and food sovereignty in the Middle East. This is where their interest in agriculture began. "At that time, we immersed ourselves in influential books about agroecology, written by pioneers like Eliot Coleman and Masanobu Fukuoka. We realized that we too wanted to start a farm, to be a part of this global movement for greener food production," they explain. Together, they explored farms in Brazil, the United States, and Quebec, eventually settling on a 12-acre plot in Saint-Christoly-Médoc, in 2018.

First steps as WWOOFers in the Americas

After leaving Israel to return to France, in 2015, Suzanne and Antoine decided to get an education in farming and enrolled in an online horticulture BTS, or technical diploma, with the ESA Angers school of agriculture. "We took this opportunity to spend eleven months roaming, visiting farms abroad. Our goal was to figure out, by the end of the year, if we could start a microfarm." Their WWOOFing journey began with running a farm in Brazil for three months, in the mountains surrounding the capital, Sao Paulo. "Managing the farm was a bit of a baptism by fire because the bosses left just after we arrived and we were on our own! In the field, we applied the recommendations we'd read about in our reference books, we repaired what we could, and we experimented. At the time, we always had Jean-Martin's book on hand; it presented a model for the agricultural transition that wasn't self-sacrificing, and it showed that, to be profitable and to be paid well, farmers needed a business model. His book gave us the confidence to start our own

FARM STATS: LES JARDINS DE LA BANQUISE

SIZE OF THE FARM

12 acres, including 8.5 acres of grassland for mulching (mowed in September)
Land for market gardening: 1 acre of farmed beds, including 1,000 ft² of high tunnels and a 2,000 ft² nursery

SALES

2022: €82,500 (US$89,000) (projection)
2021: €66,500 (US$72,000)
2020: €56,000 (US$60,500)
2019: €32,000 (US$34,500)

YEARS OF OPERATION

Since 2019

MARKETING & DISTRIBUTION

CSA boxes: 15%
Market: 45% (only one per week from
   March 15 to December 1)
On-farm sales: 15%
Restaurants: 25%

VARIETAL DIVERSITY

Open-pollinated and heirloom seeds only (no F1 hybrids):
45 vegetable crops and more than 350 varieties plus many varieties of annual and perennial flowers
Best crops: tomato, mesclun, scallion, flower bouquets
Rare vegetables: ginger and edamame

NUMBER OF EMPLOYEES & HOURS WORKED

3.2 FTE spread over the year:
• 2 market gardeners, 40 hours/week
• 1 caregiver, 25 hours/week
• 1 apprentice, 35 hours/week (from March to September)
• 1 seasonal worker, 35 hours/week (from March to September)

INVESTMENTS (EXCL. LAND & REAL ESTATE)

Total invested since the beginning: €136,000 (US$147,500)
Year 1: €44,000 (US$47,600) for greenhouse, irrigation system, including well drilling, walk-behind tractor, truck, tools, connecting to the power grid, connecting to the potable water network, earthworks, drainage
Year 2: €43,000 (US$46,500) for access road, earthworks, irrigation, tunnels, tractor
Year 3: €31,000 (US$33,500) for shelters, access, irrigation, electricity
Year 4 (in progress): €16,100 (US$17,500), broken down into €14,000 (US$15,000) for a new truck, €1,100 (US$1,200) for irrigation, €1,000 (US$1,100) for tools
For some of our investments (out of €110,000 [$120,000]), we received €44,000 (US$47,500) (40%) from the local and regional governments.

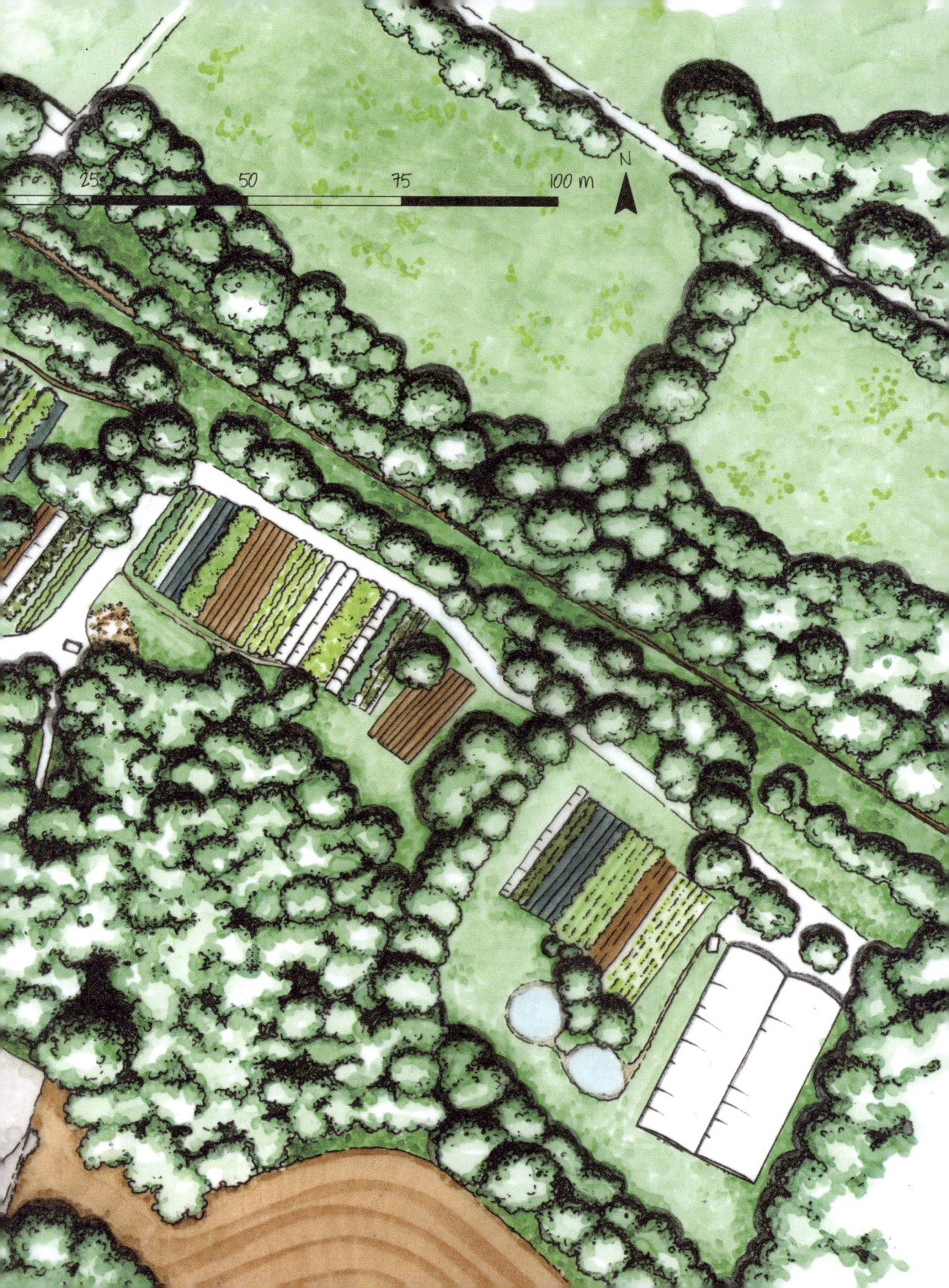

25   50   75   100 m

N

Plan to grow as many green manures as you can. Secure several short internships and work one or two full seasons with multiple market gardens. Go out and seek advice, ask neighbors about the local soil quality and whether there is standing water sometimes, to figure out whether earthworks will be needed. Keep some shaded areas, and plant trees in anticipation of hot weather. Use machines or small tractors to spread compost and straw. Lastly, if we had to start over or replicate a farm project, we would look for land that comes with good access and buildings, to be more efficient from the beginning and save our bodies and brains the effort.

project because it addresses the most important considerations, and we realized that growers could actually make a good living off farming. So we decided to make our way north, to the United States and Quebec, to get a closer look at what people were doing up there." They went on to spend a month at La Grelinette and a month and a half at Eliot Coleman's Four Season Farm in Maine, and then worked an entire summer at Ferme aux Petits Oignons. "Aux Petits Oignons is an 11-acre mechanized organic farm that feeds more than 500 families, through organic CSA boxes and one market northwest of Montréal. We learned a lot about the effectiveness of teamwork, but we also realized that we wanted to operate a non-mechanized and small-scale farm, like Eliot Coleman and Jean-Martin. After spending time on these three farms, we had everything we needed to get started."

Choosing land and starting up in the Médoc region

Back in France, they set out to find land that was close to both their hometowns, Poitiers for Suzanne and Clermont-Ferrand for Antoine. They narrowed their search down to the area surrounding Bordeaux, which is a hub with significant opportunities for sales to consumers and restaurants. Plus, it's near the ocean, "for surfing!" "We met Brice Alban, the only organic grape producer in the village, who was renting land from the Saint-Christoly-Médoc municipality. The mayor quickly offered to give us a tour of the available municipal land. One month later, we signed a lease, despite the fact that these plots are located in an official flood zone.

The land does provide some advantages, in that it comes with a nine-year rural lease, which provides security, and it is affordable (at €360 [US$390] per year for 12 acres). "These are old grasslands that were never farmed, and because our location is at the edge of an estuary, the soil is nutrient rich. Analyses have shown that it has a high clay content (37% clay, 2% sand, and the rest is silt). But it is different from what we've worked with elsewhere as the soil moisture is quite high and the land drains poorly."

Suzanne and Antoine live 20 minutes from their farm, which they registered as a GAEC (a French collective farming structure) in February 2019. Together, they were able to access a combined €28,000 (US$30,500) in grants for young farmers. "By each contributing €8,000 (US$8,500) from the start, we were able to fund our initial investments: one tunnel, the irrigation system and well-drilling services, a walk-behind tractor, a connection to the power grid and the potable water network, earthworks, and the purchase of tools and a truck."

Learning to work with clay soil

To loosen the ground and prepare beds, Suzanne and Antoine initially used the beloved BCS walk-behind tractor. "This turned out to be unsuitable for soil with a high clay content as the harrow tends to create a plow pan. Although we do follow most of the principles in the biointensive system, we realized that we

had to apply the model with some flexibility. We use every available square foot, grow different crops together, apply a rotation system that includes cover crops, and optimize our bed space. But we still have to work our soil according to the Maraîchage sur Sol Vivant (MSV, or market gardening on living soil) principles. This means no tilling and measuring inputs of organic matter (compost, cow manure, and green manure), so that the soil will develop a good structure, drain more effectively, and improve over time. We mulch 80% of the beds with hay harvested from our own 8.5-acre grassland. We combine this with compost and manure for variety and to balance nitrogen and carbon inputs. In our aisles, we also add ramial chipped wood (RCW) supplied by the municipality. This keeps weeds in check and absorbs surface water, especially in field blocks irrigated with a sprinkler system."

To manage the effects of soil moisture, which makes the clay contract or shrink, Antoine and Suzanne also had to invest in improving the drainage on some of their land. They did this at the end of their first season, with a mini excavator. "In the second year, we opted for a more beneficial system on the rest of our land, and built in swales. These small canals collect surface water and guide it towards the retention basin. We can then use water in this basin, as well as the well we had dug in 2019, to irrigate our crops."

Green manures and flowers

The methodology that Suzanne and Antoine chose largely relies on cover crops that are sown in the fall, blanket most of their fields all winter, and are mown in late March or early April. "Green manures allow us to compete with weeds, they improve soil structure, and they provide life and fertility. When we mow them, we leave the residues on the beds. The idea is to generate organic matter on-site and avoid having to source it from elsewhere," they explain. "Before starting the season, in February and March, we cover these beds with black silage tarps to warm up our wet soil, until it's time for spring and summer planting." Over the winter, one part of the farm is covered in green manure and 4 blocks are mulched: "In early spring, we move the straw into the aisles, spread the compost, and sow carrots, radishes, and turnips in these beds."

Les Jardins de la Banquise also grows flowers according to practices inspired by the Floramama flower farm, in Quebec, which is a part of the *Slow Flowers* movement. "We love the variety, color, and beauty they bring to our fields. I'm in charge of this aspect of our operations, and that means each person can have their own responsibilities, their own space," says Suzanne.

Marketing and sales

The market in the village of Saint-Vivien-de-Médoc is now one of the main outlets for their produce. It wasn't easy to get a spot at the market, given that there were already a dozen vegetable sellers, three of whom were farmers. "We get along with the mayor, and thanks to his support, we were able to get our stall set up. This is an asset for us because this market is

---

THE BIOINTENSIVE SYSTEM ACCORDING TO SUZANNE AND ANTOINE

With biointensive market gardening, you can farm on a smaller plot, which means you are less subject to constraints related to land. Crop diversity gives growers the opportunity to test a variety of vegetables and introduce their customers to new produce, all while avoiding the risk of incurring significant losses if one crop is affected by a weather event. Once market gardeners implement this system, it can provide some everyday comfort: in the fourth year, we were able to free up more time to do other things like organizing our personal life and taking courses on other topics.

sonal produce from April to December. We sell some CSA farm boxes, 20 boxes for a 30-week season, make some sales on the farm, and also sell seedlings and herbs from mid-March to June."

To increase profits every year, Suzanne and Antoine corrected and tweaked their plans based on their yields and demand, even when that meant offering fewer vegetables. So far, they've reinvested all profits back into the farm, and have yet to pay themselves; instead, they have been living off their savings and government support. They had planned to start paying themselves a salary in 2023. "With this approach, we were able to take out fewer loans, and today we have a system that is comprehensive and effective. We don't have kids yet, so our responsibilities are minimal. Since we don't live on the farm, we plan to build a small structure that will house a kitchen, office, and break room."

Since flooding is an ever-present risk on their land, they let the field rest from the end of November to March, which gives them roughly six weeks off between December and January. Fortunately, our two intrepid farmers have not had to contend with any flooding since they started the business. "But it is kind of this Sword of Damocles, threatening everything that we invest in, and it forces us to shut the whole operation down for winter." With their usual joy and laughter, they add, "So, at the end of the day, if the water rises, we'll become fishermen!"

renowned in our area. It has helped us build a reputation. We visited chefs, tweaked things as we went along, focusing on the people with whom we worked well and had a friendly relationship. We are fortunate to collaborate with La Fleur au Fusil, one of the best restaurants in the region, in Saint-Vivien-de-Médoc, as well as four other restaurants, bringing them sea-

KEY DATES

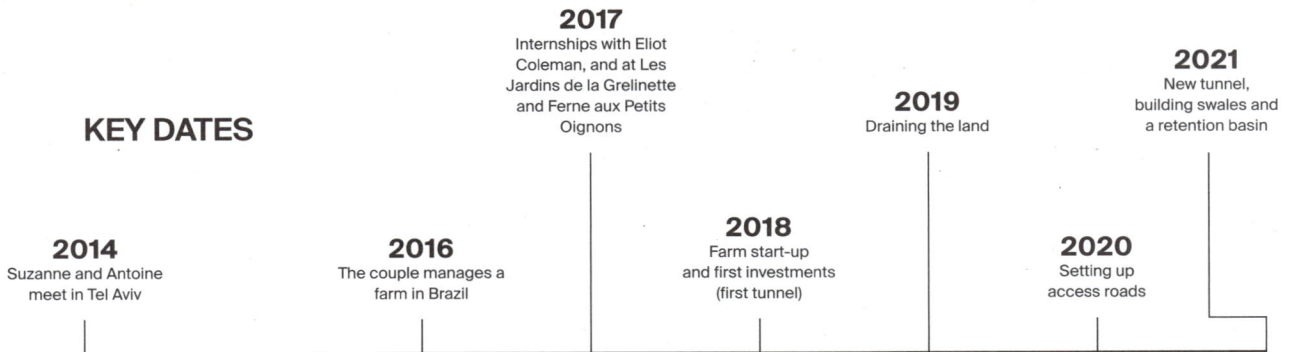

2014
Suzanne and Antoine meet in Tel Aviv

2016
The couple manages a farm in Brazil

2017
Internships with Eliot Coleman, and at Les Jardins de la Grelinette and Ferne aux Petits Oignons

2018
Farm start-up and first investments (first tunnel)

2019
Draining the land

2020
Setting up access roads

2021
New tunnel, building swales and a retention basin

Azienda Agricola Foradori

Myrtha Zierock

ITALY

Myrtha Zierock grew up surrounded by vineyards, on the Foradori estate. Located in Mezzolombardo, in the Dolomites in northern Italy, it was founded by her great-grandfather in 1935, and has now been managed by her family for four generations. In 2002, Myrtha's mother, Elisabetta, led the vineyard through a transition into biodynamic practices. In the world of wine, Elisabetta is a true legend as she worked to restore the genetic diversity of the teroldego grape, also known as "gold of Tyrol." Today, Myrtha, her mother, and her brothers co-own this 74-acre operation, where they generate an average of 180,000 bottles per year, exported to 35 countries around the world. Since 2019, Myrtha has also been developing her market garden on 0.3 acres. She grows top-quality vegetables, meant to be enjoyed with a glass of local wine.

From Germany to Quebec via Oregon

After studying environmental science at the University of Freiburg, in Germany—with a focus on understanding and protecting ecosystems—Myrtha flew to Oregon, USA, for a one-year exchange. "I spent a few months applying my knowledge at Ayers Creek Farm, and this included seed selection," she says. "I started to think about food quality, and realized I could see myself growing tomatoes more than grapes. I fell in love with vegetables and their diversity, despite the effort it takes to grow them. At a conference in February 2015, Myrtha came across a small-scale farming model that was both environmentally friendly and profitable; it was presented by Jean-Martin Fortier, who would later become her mentor. Over the next three years, she learned bio-intensive techniques and became familiar with the principles of the system, through a 5-week internship at La Grelinette and two seasons as an employee at Ferme des Quatre-Temps. "I learned how to grow crops and manage a team, how to create a production plan, and how to work with crop successions. That's when I realized that I could take this model and apply it right in the middle of my vineyard, that a market gardening project could bring diversity to Foradori."

FARM STATS: AZIENDA AGRICOLA FORADORI

SIZE OF THE FARM
74-acre vineyard and 0.34-acre market garden containing 0.26 acres of beds:
Main plot: 0.2 acres (4 × 9-bed blocks, each bed is 2.6 × 50 ft, for a total of 0.11 acres of growing surface)
Beds between rows of vines: 0.16 acres total

SALES
2021: €25,000 (US$27,000)
2020: €25,000 (US$27,000)
2019: €10,000 (US$10,800)

YEARS OF OPERATION
Since 2019

MARKETING & DISTRIBUTION
Direct sales on the farm twice a week (45%), selling to restaurants (40%), lacto-fermented vegetables (10%), catering and events (5%)

VARIETAL DIVERSITY & BEST CROPS
26 kinds of vegetables and 60 varieties
Best crops: salads, tomatoes, Swiss chard, carrots, Asian mesclun

NUMBER OF EMPLOYEES & HOURS WORKED
2 FTE spread over the year:
- 1 farm manager at 15 hours/week, year-round
- 1 person at 40 hours/week, from March to mid-December
- 1 person at 20 hours/week from March to October
Payroll for 2 employees, over one year: €20,000 (US$21,500)

INVESTMENTS (EXCL. LAND & REAL ESTATE)
€25,000 (US$27,000) in the last 3 years, excluding seasonal purchases (seeds, amendments, packaging, etc.)

Returning to the family vineyard in Italy

In December 2018, Myrtha returned to the Dolomites and started her market garden, following the biodynamic methods that had been applied on her family's vineyard for over 15 years. "After spending so many years abroad, it was hard for me to establish my own space on the vineyard, with all the ongoing business operations, such as welcoming customers and hosting visits. The first year, I worked with an intern to see what I could grow, to see how far I could get in establishing this project, and at what scale. I also tried to figure out how this dynamic would line up with operations at the vineyard. So my market garden was initially a small start-up within the estate. I was fortunate to get a €20,000 (US$21,500) budget from the family business for initial investments like building a wash station and buying tools, a mesclun harvester, caterpillar tunnels, insect netting, and a walk-behind tractor. I was also able to use existing equipment and infrastructure like the irrigation system and the cellar."

Since land is particularly expensive in Myrtha's region, her challenge was to set up vegetable crops in areas not used for vines, and to optimize these new plots. "We are in a valley surrounded by mountains, 820 feet above sea level, on a sandy-loam soil. With a walk-behind tractor, I set up my main garden on 0.2 acres of land that gets plenty of sun. I divided the plot into 4 blocks, each containing 9 50-foot beds, so that I could rotate between vegetable crops and winter cover crops. This is where I grow my spring and summer vegetables (carrots, radishes, fennel, Asian mesclun, tomatoes, herbs, Swiss chard, and zucchini). I also use the land in between rows of vines because our vineyard follows the pergola Trentina system, which is adapted to our local altitude and topography. It means we maintain a 20-foot space between rows, where I've set up 7 beds, each 5 feet wide and 200 feet long, for a total of roughly 0.3 acres." Myrtha amends the soil by consistently adding compost made from winemaking waste material, which is rich in soil-stimulating microorganisms. "In April, we remove the cover crops growing between the rows of vines, and then we don't seed anything there until August because tractors will be passing through to tend to the vines. Then, in the first week of August, we plant fall and winter vegetables, mainly into a woven ground cover to keep the weeds in check. From the end of September until Christmas, we harvest garlic, Swiss chard, kale, winter radish, and daikon radish." Having so little space to work with, Myrtha can't build greenhouses, but she has managed to squeeze in 100-foot-long caterpillar tunnels that get moved at the start of every new season. "We set them up in the fall, to shelter 4 beds of lettuce, Asian mesclun, tatsoi, radish, and turnip. I have another plot of land, 3.7 acres, about 40 minutes from the vineyard where I rotate onions and garlic and have planted fruit trees." As for seedlings, she buys them from a local organic nursery.

A farm centered around fall and winter vegetables

Although Italian winters are mild, Myrtha's spring crops are slowed down by the surrounding mountains that block some of the afternoon sunlight. "We lose the light quite early, and if it hasn't been too cold, plants like kale and Chinese cabbage that were seeded at the

---

THE BIOINTENSIVE SYSTEM, ACCORDING TO MYRTHA

With this model, you can grow a decent amount of vegetables on a small piece of land, as long as you develop a solid farm plan and don't have too many issues related to climate. The term "biointensive" can be misleading, because the term "intensive" here is different than what it would mean in conventional farming. It's about fostering a more intense soil health.

end of summer will start growing back the end of February or early March." The climate and certain regional specificities tend to favor fall and winter vegetables, grown from October to March: "Those are the crops that work best here. In the summer, people go on vacation, and many of them also grow summer vegetables in their own garden. But from November onward, they aren't harvesting anymore, so customers are ready to eat our produce again, once it gets harder to grow crops outdoors." So Myrtha has focused on a range of varieties that are better suited to this growing season: daikon radish, turnip, chicory, red cabbage, romanesco cauliflower, pac choi, mustard-leaf mesclun, onion, garlic, leek, Swiss chard, and local varieties like the Brocolo Padovano leaf broccoli and Tuscan kale. She also grows crops that can be stored over winter: carrots, beets, and squash.

A diversified multifaceted farm

As both a farmer and a single mother, Myrtha needs to schedule her time carefully so she can focus on her family life. "I hired a full-time employee, and we plan out the week together every Monday. I also hire a seasonal worker from May to October. I try to stick to a 40-hour workweek, so I can stop at 5 p.m. and take care of my two-year-old son. My time is split about 50/50 between managing certain operations on the vineyard and growing vegetables.

To recover the funds I invested and to pay employees, I estimated that sales had to total €40,000 (US$43,500) (my salary is paid outside the project because I own shares in the vineyard). The year my son was born, it became clear that I wouldn't be able to pour the same energy into the project and that I wasn't going to see great results."

In 2021, Myrtha brought in €25,000 (US$27,000) in sales, the result of a precise plan that covered all steps, from seeding through to harvesting, and optimized space, energy, and costs. In terms of marketing, half of the produce goes to restaurants, while the rest is sold directly to consumers on the estate. "Initially, my business model was primarily based on sales to restaurants, but when COVID hit, I decided to start selling produce at the vineyard. I do a lot of planning during the winter, using data from the previous year. The hard part is really understanding your local market, especially in Italy, where conventional growers can offer high-quality fruits and vegetables. I need to be able to tell my customers why they should pay more for my carrots. This is one of the limitations that some young market gardeners face; they know how to grow food but often don't know how to sell it. One thing's for sure: in direct sales, if the quality is there, don't sell it for less than at the local market."

Myrtha has managed to grow vegetables that her customers appreciate just as much as her family's wine. Her project can serve as a model and open up a world of possibilities for wineries looking to diversify.

MYRTHA'S ADVICE

In those first five years, you must be gentle with yourself and don't slave away, trying to do everything on your own.
You have to be able to manage a range of tasks and keep your personal life and work life separate; you also have to know how to stop. I think that's the most valuable lesson, but it also requires careful planning.

If, like me, you have quite a small piece of land, you'll need to consider keeping some space free when crop planning. This will provide flexibility and opportunities to adapt if you face unexpected weather events. In fact, I find it can be quite stressful to leave very little time and space in your crop rotation plan.
At the start, don't try to grow every

vegetable right away, because it's just not possible. You'll be better off focusing on 10 to 15 vegetables, growing them well, and finding the right markets, which depend on your region and context. Opt for something like mesclun, which is popular with restaurants.

KEY DATES

2014–2015
Studying environmental
science in Freiburg,
Germany, with a one-year
exchange at Oregon
State University, USA

2015
Myrtha meets
Jean-Martin Fortier
at a conference

2016
Internship at
Les Jardins de la
Grelinette, Quebec

2017–2018
Employee for two
seasons at Ferme des
Quatre-Temps,
in Hemmingford

2018
Return to the Foradori
family vineyard,
in the Dolomites

2019
Starting up the
market garden

Wilmars Gaerten

Maria Gimenez
GERMANY

Maria Gimenez was born in Bonn in 1981, to a Spanish father and German mother. Since 2019, she's been using her market garden, Wilmars Gaerten, to cultivate the utopia of a sustainable world. Through the farm, located just 25 miles south of Berlin, Germany, she has shown that you can grow beautiful vegetables while still earning a decent living. Before becoming a farmer, Maria was a painter with a thriving career. Then one day, she walked away from it all to start a new adventure. She had been considering this change for some time. "I used to spend my summers in southern Spain when I was a kid," she recalls. "You could see how industrial agriculture was destroying the country. It's become a desert, nothing grows there. I have six kids now, and faced with the reality of a burning planet, I just couldn't see myself painting a world that was collapsing, to reach only an elite group of art lovers and intellectuals. I had to act, question the system, repair what we'd destroyed, and start a revolution!" She made the transition in 2017, when she traded the Berlin art scene for humus, bees, and a life spent outdoors.

A market garden within a landscaped park

Her partner's father helped her make the first move when he offered Maria some land near a landscaped park on his 900-acre estate, mostly made up of grasslands and forests. He set only one condition: she'd have to do something useful with it. Maria leaped at this opportunity and moved her family to Märkisch Wilmersdorf, in Brandenburg. This region is the driest in the country. It gets almost as little rain as southern Spain, and it has been shaped by monoculture, the use of pesticides, and increasingly severe droughts.

Maria chose to name her garden after Wilmersdorf, a town whose original name meant "great will." It likely required a great deal of willpower to transform her sandy ground into nourishing and productive soil over the course of four years. "I had already begun to learn about sustainable farming systems and agroforestry, but since I wasn't from the agricultural community, I knew nothing about working the land."

FARM STATS: WILMARS GAERTEN

SIZE OF THE FARM

1.5 acres of farmed land
The farm has been designed to eventually reach a total of 12 acres.

SALES

2022: €160,000 (US$173,000)
2021: €75,000 (US$81,000)

YEARS OF OPERATION

Since 2021

MARKETING & DISTRIBUTION

Direct sales: 4 weekly markets in Berlin and 1 on the farm
Selling to 20 restaurants, including Michelin-starred Nobelhart & Schmutzig and Ernst

VARIETAL DIVERSITY & BEST CROPS

40 types of vegetables with over 150 different varieties
Best crops: carrots, beets, tomatoes, and beans

NUMBER OF EMPLOYEES & HOURS WORKED

4.8 FTE spread over the year:
• 3 permanent employees
• 4 interns (3 to 6 months)
• 1 accountant
Total of 230 hours/week €96,000 (US$105,000) payroll in 2021, and €134,000 (US$145,500) in 2022

INVESTMENTS (EXCL. LAND AND REAL ESTATE)

Total invested since the beginning: €125,080 (US$135,500)
2021: €92,840 (US$100,500) for the entire market garden project and all work done on the 12 acres (preparing the soil, building greenhouses, setting up electricity, fences, irrigation systems, purchasing tools and seeds)
2022: €32,240 (US$35,000) for pest control, organic fertilization, distribution and sales logistics, software, extending the irrigation system, and adding beds

Maria taught herself by reading books about new farming systems, and she developed a passion for regenerative and biointensive agricultural practices. She came across the methods that Jean-Martin and Maude-Hélène were using at La Grelinette, and she used their operation as a model for her start-up phase. "I began quite small, on 1.2 acres, with a student who was helping me figure out the amount of money and the number of people I'd need to farm 12 acres. Today, 5 of us work on 1.5 acres, split into 7 field blocks, each containing 30 130-foot-long beds. We grow more than 40 types of vegetables, about 150 different varieties, as well as a wide range of herbs and edible flowers, all without machines or pesticides."

Gradually, and without any tilling, Maria was able to restore the soil and make it productive by adding a 6-inch layer of compost to every bed. The improved humus content helped loosen the soil deeper in the ground, resulting in better moisture distribution. She is also experimenting with new water retention basins. "We use the soil, but we do our best to give back to it. For the past two years, we've been making our own compost from our own organic matter, such as alfalfa, clover, dead leaves, wood chips, and manure from cows and horses. We try to think of new practices like using as little plastic as possible, for instance. By starting our own seedlings in our nursery, we have control over everything we put into the soil, and we work mostly with open-pollinated and heirloom seeds to break free from the supremacy of big seed companies."

Toward a regenerative form of agriculture

On the farm, Maria has implemented a wide range of concepts from regenerative agriculture. Some plots of land are dedicated to grazing for their 33 dairy cows. They also installed 70 hives according to the Warré organic beekeeping method, so the bees can live in an environment that is similar to their natural state. "The idea behind Wilmars Gaerten is to create a resilient and edible landscape that shows many different regenerative approaches to growing food—not just for us humans but for all living beings around us. So far, we've planted over 170 acres of different agroforestry systems, which includes more than 200,000 trees and over 35 varieties. Our livestock grazes both on cultivated land, between the trees in our agroforestry systems, and on grasslands. Before starting our first market gardening season, we planted 5,000 poplars to counter the prevailing winds and ran chickens in between the rows of trees. We also planted more than 2,000 fruit trees and various berry bushes. At this point, in the vegetable garden, we've planted 50 fruit trees between the field blocks. Beneath the trees, we grow vegetables, nuts, herbs, and edible flowers. We want to show that it is possible to feed many people without needing to destroy nature, by developing regenerative agricultural practices, because soil health is the key to our future." The farm currently offers workshops for schoolchildren and students, and will soon extend that to anyone interested in regenerative practices, whether that means learning how to grow vegetables in a market garden, or beekeeping.

In the summer, they also host chefs who teach cooking classes and prepare meals.

THE BIOINTENSIVE SYSTEM, ACCORDING TO MARIA

The future is in the hands of small farmers. It's a lot of work, with so many problems to solve, and it's not all rosy.

Sometimes you'll make mistakes. But it's a wonderful, rewarding job!

And on our biointensive farm, we see only happy faces!

Marketing and distribution in Berlin

It wasn't hard to find outlets for Wilmars Gaerten's first harvests. Within a few months, Maria had made a name for herself in Berlin markets with her elegant vegetable stand. Berlin's most famous kitchens were quick to embrace her high-quality produce, which is exclusively grown by hand and with natural practices. "We had a great start to the first season, so we were able to sell our vegetables at two markets in downtown Berlin and to renowned gourmet restaurants. That resulted in €75,000 (US$81,000) in sales. We had actually projected €140,000 (US$150,500) in revenue for that first season, but we had to contend with several challenges related to weather, pests, and probably a lack of experience." At four weekly markets in Berlin, they sell their produce: initially fruits and vegetables, and now also honey, eggs, and bread. In 2022, these sales totalled €160,000 (US$173,000).

Through direct sales, Maria also became aware of the political impact of organic local farming. "Our vegetables are both beautiful and tasty, they exude a positive energy. At the farm stand, people can tell that we cared for these crops throughout the growing process. This is in stark contrast to the produce sold in supermarkets, which are more like "dead" vegetables. By creating a nurturing landscape with my team, just like I used to create art, I've come to realize that we all have the capacity to create the reality in which we want to live. Our customers are also a part of this creative revolution, we're making this change together.

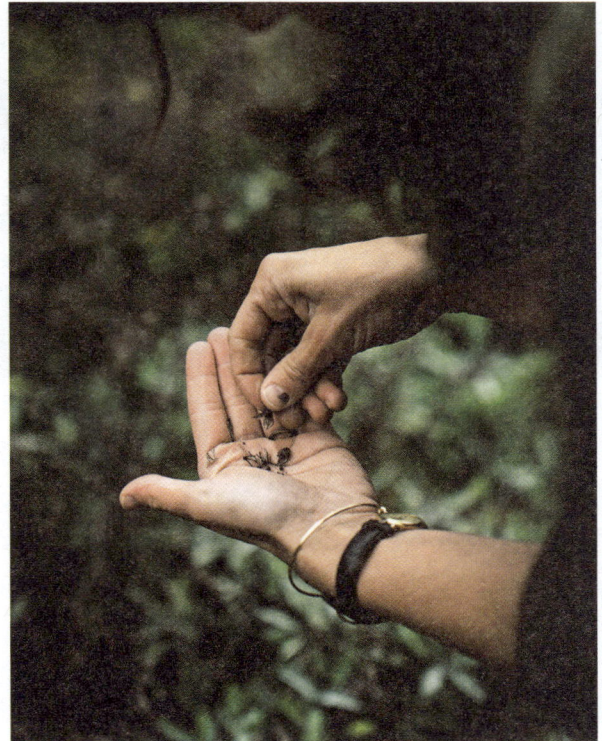

KEY DATES

2017
Maria quits painting and decides to start a regenerative farm on her father-in-law's estate in Brandenburg

2018
Maria plants 200,000 trees, according to Keyline design techniques, and sets up hives

2019
First harvests, sales at markets and to restaurants

2020
Pause during COVID

2021
First real season

MARIA'S ADVICE

Use a model but don't follow it religiously; be flexible, rely on your creativity, and trust your intuition. I found help, knowledge, and inspiration in books and through the experiences of other farms. But you need to study your own land and environment to understand and to adapt.

I want everyone to remember that each of us was born with a creative energy—we are able to create the world in which we want to live. We don't have to accept the destruction of our only home, our planet. We have to come together and change things! You'll have to face so many unpredictable situations, but don't give up!

La Fermette

Annie-Claude Lauzon
and Justine Chouinard
CANADA

Annie-Claude Lauzon and Justine Chouinard are partners in both life and work, and neither of them grew up in the country. They hail from Montréal and the city's suburbs on the south shore, respectively. Justine studied visual arts and art history, while Annie-Claude specialized in environmental science. While in school, both women were involved in various urban agriculture and beekeeping projects. "Working with these organizations got us thinking about the social and environmental impacts of the current food system. It made us want to be a part of the movement for ecological and local agriculture," explains Annie-Claude. That's how, in 2012, they both ended up with an internship on a peri-urban organic vegetable farm run by Santropol Roulant, an organization that promotes community food security. They then worked at D3-Pierres, a farm providing social integration services, also on the island of Montréal.

"Through these experiences, we became even more interested in starting our own farm," explains Annie-Claude. "The biointensive model appealed to us because of its scale, accessibility, and farming techniques. We felt like we needed to learn more about this model, and that's how we ended up on the Ferme des Quatre-Temps team. It's also where we met Jean-Martin and the people who would eventually become our business partners."

A formative experience at Ferme des Quatre-Temps

Working at Ferme des Quatre-Temps while it was being established was a formative experience in their agricultural journey. "We were a part of a farm's first year of operation. Going through the start-up phase was very rewarding for us, because this was an opportunity to get involved and see, firsthand, how the project evolved: implementing the production plan, marketing to restaurants, getting a farmers' market set up."

In their two seasons at Ferme des Quatre-Temps, they learned practical skills and crop production, while developing their ability to manage a team. "We each had our own responsibilities; Justine was in charge of the nursery, and I ran the harvests and planting. We were tasked with sticking to the production schedule and managing the team when transplanting and direct seeding in the field."

This experience played a key role in their journey as it gave them the necessary tools to start their own farm and it made them want to work hand in hand with restaurateurs.

Once their contract was over, Justine and Annie-Claude looked for a small piece of land near Montréal, "which is no simple task in Quebec, where farmland is typically sold in large lots. Justine and Annie-Claude also shared the ups and downs of La Fermette's

FARM STATS: LA FERMETTE

SIZE OF THE FARM

Total land area: 5 acres
Farmed land: 1.6 acres
Heated greenhouses: 0.2 acres (8,000 ft²)
Unheated greenhouses: 0.06 acres (2,700 ft²)
2 × 500-foot tunnels
Between 18 and 20 blocks, each with 12 × 50-foot beds;
total of approx. 240 beds in the field

SALES

2021: Can$235,000 (US$174,000)
2020: Can$225,000 (US$166,500)
2019: Can$165,000 (US$122,000)
2018: Can$100,000 (US$74,000)

YEARS OF OPERATION

Since 2018

MARKETING & DISTRIBUTION

Direct sales at farmers' markets: 20%
Preorders and online marketplace: 45%, with 3 delivery
locations: 1 on the farm and 2 in Montréal (Old Port of
Montréal and La Buvette Chez Simone)
Sales to restaurants and grocery stores: 35%
(50% before the pandemic)

VARIETAL DIVERSITY & BEST VEGETABLES

55 vegetable varieties
Top 5 vegetables: greenhouse tomatoes and cherry
tomatoes, Lebanese cucumbers (grown in greenhouses),
mesclun, Nantes carrots, radishes

NUMBER OF EMPLOYEES & HOURS WORKED

6 FTE spread over the year:
• 2 market gardeners + 4 employees in 2022 (3 in 2021)
Total of 7,000 hours in 2021 including Annie-Claude and
Justine (roughly 3,500 hours if excluding them)

INVESTMENTS (EXCL. LAND & REAL ESTATE)

Total invested since the beginning: Can$185,000
(US$135,000)
2018: Can$75,000 (US$55,500)
2019: Can$25,000 (US$18,500)
2020: Can$50,000 (US$37,000)
2021: Can$35,000 (US$26,000)

start-up on TV, when they were featured in the second season of Les Fermiers. "It can be stressful to be filmed during your farm's start-up phase, when everything is new and you're dealing with the unexpected. But this media coverage was an opportunity for us, because it got us out in front of the public," they admit.

A farm that partners with restaurants

Justine and Annie-Claude were also fortunate to have met the owners of Café Parvis and La Buvette Chez Simone, two Montréal restaurants. They had been wanting to help start a farm and were looking for collaborators who could make this project a reality. Meanwhile, Justine and Annie-Claude had been looking for partners to join them in their farm's start-up. And thus, they partnered with the two restaurateurs to create what would eventually become La Fermette.

But first, they had to figure out where to start this future market garden. After doing some research, Justine and Annie-Claude found a 5-acre lot in Hemmingford, near Ferme des Quatre-Temps, "an old pasture with a small barn, a house, a stream, a well, and a small woodlot." The farm is one hour, at most, from Montréal and the majority of their sales outlets, which is a major advantage. Their location does come with significant challenges though, like drainage issues and the need for rock picking: "Our soil is rocky, and we had to remove a lot by hand to establish the beds. We repurposed the rocks for things like pinning down our silage tarps." To get started, Justine and Annie-Claude secured a Can$85,000 (US$63,000) loan for their initial investments, through the Financière agricole du Québec, and subsidies for emerging farmers in Quebec. "With these funds, we were able to finance our first greenhouse, a vehicle, a walk-behind tractor, many tools, silage tarps, and floating row covers… So the first step was to build the greenhouse. We were able to rely on valuable help from our friends, family, and team members at our partner restaurants. The greenhouse is automated and heated, to help us extend the season at both ends. It's an important asset for us because it means we can plant tomatoes at the beginning of April and start harvesting in June."

Justine and Annie-Claude were also quick to design a permanent system of beds, each 50 feet long, rather than the 100 feet recommended by Jean-Martin. "This length is a better fit for our farm's needs, and it still allows us to grow a wide variety of crops. We started out as a team of two, and it was just easier to lay out nets and silage tarps on this smaller area. It also helps us to keep morale high when, for example, we have to weed four beds of carrots!" Over the following years, they built three more greenhouses and two caterpillar tunnels that they move over the course of each season. They also dug a pond to provide an additional water source.

ADVICE FROM JUSTINE AND ANNIE-CLAUDE

Before starting your own operation, you need to learn the trade by working at least one season (or two, if you can) on a farm that inspires you. In this profession, you really learn from experience, by practicing each step: seeding, transplanting, crop maintenance. Through this work, you can acquire physical skills and field experience, which will help avoid making too many mistakes when starting up your own farm. It's also an opportunity to learn how to work with nature and living things, which are vulnerable to hazards that are outside of your control (disease, pests, and weather events).

Our fast start-up didn't give us time to properly prepare the field blocks. As a result, we had to deal with drainage problems along the way. You'll be better off taking a year to work your soil properly and set up all your infrastructure and systems.

As a market gardener and business owner, you have to budget time for daily problem-solving (mechanical and technical, human, production). You absolutely need to strike a balance between the three cornerstones of any market garden: harvests, team size, and sales. If you're growing too little food, the farm won't be profitable. If your team is too small, people will feel stressed. If your vegetables are beautiful but you don't have a market, they'll go to waste.

BUVETTE
×
FERMETTE

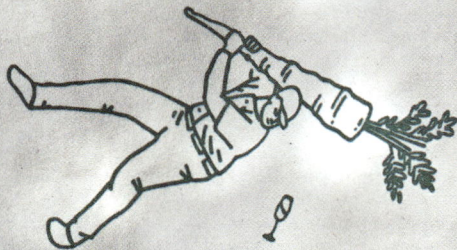

In 2020, when COVID hit, they joined a new project designed to let customers preorder vegetables through an online platform. "It was a great way to assess how we'd sell our produce in this new context, and to adapt. With this system, we started harvesting vegetables to order. We love the flexibility that this platform provides, both for us and for members. Plus, it sets us apart and reduces the time we spend managing unsold goods. And our prices are similar to those in grocery stores. Our customers are aware that by buying our vegetables, they are contributing to a movement for a different and sustainable agriculture.

From mid-May to early November, their farm now supplies their two partners as well as other restaurants and consumers. "It's the equivalent of about 200 boxes and the semi-wholesale of roughly 15 select vegetable crops. With some vegetables, like lettuce, tomato, radish, cucumber, white turnip, and Nantes carrots, we grow large quantities to supply restaurants. However, others are seasonal and just for consumers, including new potatoes, beans, sweet peas, and broccolini."

Life at La Fermette follows seasonal rhythms: it's intense in the summer during the harvest and sales season, then it's calmer towards the end of the fall and start of winter. "Every year, the team that supports us gets bigger, and we delegate more and more tasks to the people we work with. In the high season, this means we've been able to take Sundays off and reduce, a little, our hours of work. We also take just over one month off in the winter, which gives us time to nurture our relationships and the other spheres of our lives, to pursue activities and to strive for balance with this demanding work," they explain.

In a society that is still dominated by a gendered and traditional view of the agricultural world, Justine and Annie-Claude are inspiring role models for women who want to start a profitable, ecological, and productive market garden.

THE BIOINTENSIVE SYSTEM, ACCORDING TO ANNIE-CLAUDE AND JUSTINE

When we started our farm, the biointensive microfarm model seemed accessible to us, in that it required just a small plot of land, hand tools, and minimal investments, when compared to other agricultural models. With biointensive methods, you can farm on a small scale by optimizing operations through a system designed for manual work, and you can maximize yields with crop successions and denser spacing.

Running a diversified microfarm also means tackling a broad range of daily and weekly tasks, which keeps the work stimulating throughout the season.

KEY DATES

2012–2015
Annie-Claude and Justine get involved in urban agriculture projects (collective gardens and beekeeping), as well as peri-urban farms with a social purpose

2016
First season at Ferme des Quatre-Temps, in Hemmingford, learning the biointensive system

2017
Second season at Ferme des Quatre-Temps, managing the nursery and planting, coordinating teams

2018
Starting the farm in Hemmingford, building the greenhouse in winter; start-up thanks to the help of many volunteers and the support of two partner restaurants (Café Parvis, La Buvette Chez Simone), Can$100,000 (US$74,000)

2019
Hiring a first employee, Can$165,000 (US$122,000) in sales

2020
Spike in demand for farmers' markets during COVID. Despite the complete or partial closure of several restaurants, all produce is sold, primarily at the farmers' market and thanks to a new preordering system through an online platform

La Microferme des Anges

Jean-Michel Le Guen

FRANCE

In 2018, Jean-Michel Le Guen started Microferme des Anges in Orvault, a few miles northwest of Nantes, France. On this new farm, he applied the biointensive system that he had set up on his previous operation, in Gironde, back in 2012. Jean-Michel is a former landscape gardener who converted to agriculture. He is self-taught and a hands-on kind of person. While his farm was initially a classic mechanized market garden, his instincts guided him towards a transition in which he reduced the growing area from 3.7 to 2 acres. "Coming from a professional landscaping background, I found farming hard. I wanted to set up the fields in a way that would make the work easier, so I made a few beds that were only 2.5 feet wide. These were so much more accessible than my original 4-foot beds," he explains. "It saved me a lot of time. Little did I know, my idea was in line with a method recommended by a Quebec market gardener, Jean-Martin Fortier, who had just published his first book. I read it, and watched all the videos he posted on social media. Within a year, we'd changed everything to apply his model. We increased our crop density, took good care of our soil and covered it. In some blocks, we intensified our rotation to include four crop successions. The results were phenomenal. We increased our yields, delivering vegetables to roughly 120 families. And with the profits, both my business partner and I were able to live off the farm's operations."

A small but optimized field

In 2016, Jean-Michel walked away from his first market garden and moved to Nantes to train as a brewer and then bought 1.2 acres of land to grow hops. In the end, he started another biointensive market garden there, on 0.6 acres. "Before starting the farm, I first mapped out the field blocks. I was on fallow land, with well-balanced sandy clay soil, so all I had to do was feed it enough to keep it alive." Jean-Michel's fields are on a south-facing 14% slope. Located in an old quarry, it has underlying rock that allows for no more than 12 inches of natural soil depth. "With compost inputs, roughly 65 cubic yards spread over the beds, I was able to make my soil more productive and increase the depth." There aren't any walk-behind tractors on this farm. Just "a seeder and a tilther, recommended by an American market gardener called Eliot Coleman. We work the land down to a maximum of 2 inches, to avoid destroying microbial

FARM STATS: MICROFERME DES ANGES

SIZE OF THE FARM

1.2 acres: 0.75 acres are farmed, with a 7,500 ft^2 greenhouse and a 500 ft^2 propagation greenhouse

SALES

2022: €41,000 (US$44,000)
Includes €31,000 (US$33,500) from the market garden and €10,000 (US$11,000) from a course given through the local Chamber of Agriculture
2021: €21,000 (US$22,500)

YEARS OF OPERATION

Since 2018

MARKETING & DISTRIBUTION

Sales to gourmet restaurants, 15 of which are Michelin-starred (67% of revenue)
Education (33% of revenue)

VARIETAL DIVERSITY

100 varieties, specializing in mini vegetables

NUMBER OF EMPLOYEES & HOURS WORKED

No employees (one apprentice)
40 hours, including work with the Chamber of Agriculture

INVESTMENTS (EXCL. LAND & REAL ESTATE)

€15,000 (US$16,000) + recovery of equipment from previous farm (tunnels, tarps, irrigation)
€10,000 (US$11,000) more in the last 3 years, self-financed

ADVICE FROM JEAN-MICHEL

I recommend taking the Market Gardener Masterclass because I think it's a great tool, it really provides everything you need to get started. It's hard to get access to land, so if an opportunity presents itself, you have to jump on it. After that, you'll have plenty of time to fine-tune your farm plan. I'm certain that biointensive market gardening can work anywhere, but I'd avoid settling down near vineyards because their disastrous agricultural processes affect soil health.

life and to protect the soil." In the winter, from December to February, Jean-Michel lets the fields rest, covers them with compost, and sows cover crops. "To maximize my land use, I've shaped about 100 raised beds, each 30-inches wide and 30- to 65-feet long. So that gives me about 25 1,000-square-foot blocks for crop rotations. My 10-inch aisles are smaller than the ones recommended in Jean-Martin's model (16 inches), to make the most of my small surface area. I can still get around fine, because I move sideways along the bed."

Jean-Michel, now in his 50s, also received a 20% refund from his regional government, when he invested €23,000 (US$25,000). "I also took out a €30,000 (US$32,500) loan, which I've now fully repaid. I was able to use it to drill a well, excavate a water retention basin, buy tunnels and irrigation equipment, set up a small workshop, and start constructing a building myself. From my old farm, I was able to recover some infrastructure, like a 7,500 square-foot greenhouse where I grow 95% of my seedlings, mostly heirloom varieties, no F1 hybrids. I like to look for obscure varieties, such as the Petit Moineau (Little Sparrow) cherry tomato and grape tomatoes. I work with small seed companies on some varieties, and I try to save my own seeds."

Between field blocks, he's planted about 40 fruit trees: apple, pear, peach, plum, and cherry. They're set 15 to 30 feet apart, to provide a little shade for vegetables that suffer under the summer heat and sun. "Trees are precious allies for vegetables. They offer shade, and falling leaves supply free organic matter that enhances biological life in the soil. The goal is also to enrich our soils naturally and to bolster microbial and underground life. The other advantage comes from tree roots, which loosen and aerate the earth," explains Jean-Michel. "I limit the use of irrigation at the end of a crop's life, so that the vegetables will draw up as many minerals as possible and thus take on the flavors of our terroir. Because of all these alternative farming methods, the vegetables I harvest are highly flavorful," he explains.

Capturing the hearts of Michelin-starred chefs

The best restaurants in the city do appreciate Jean-Michel's delectable produce. Since several chefs first noticed his work, mostly on Instagram, his main clientele has consisted of Michelin-starred restaurants, roughly fifteen of them. He now collaborates with seven restaurateurs in Nantes, including Michelin-starred chef Ludovic Pouzelgues at Lulu Rouget on the Island of Nantes, Jérémy Guivarch from Gwaien, and Aymeric Depogny from Abbaye de Villeneuve. "I started posting on social media in 2015. First one chef from Monaco wanted to collaborate with me, and the others followed suit. The chefs are very fond of my products, especially in Nantes, where demand is exploding, so I've even had to put some of them on a waiting list! I grow mini vegetables, and cooks can't get enough of them: colorful carrots, zucchini, fennel, leek, beets, pac choi, kohlrabi, and also cherry tomatoes, herbs, fresh chickpeas, beans, green peas…" Since 2019, Jean-Michel has been a member of the Collège Culinaire de France, a network of restaurateurs and farmers. "I work to meet the needs of chefs, especially those with whom I've been collaborating for years. Sometimes they point me towards this or that variety. These are direct conversations, without intermediaries."

A dedicated space for teaching

As the only biointensive farm in Nantes and the surrounding area, Microferme des Anges is of interest to the Pays de la Loire Chamber of Agriculture, due to the resurgence of new farmers applying to start small-scale operations. "They asked me to train young aspiring farmers by creating a 0.5-acre educational space for students on my land. The model is appealing to a growing number of candidates who have applied for government approval to start their own farm because it ticks many boxes: it's organic and it's productive. And I've been able to witness this evolution; three years ago, people were only talking about permaculture, last year it was MSV, and now this year, it's all about biointensive methods. So I provide training about MSV [Maraîchage sur Sol Vivant or market gardening on living soil] and biointensive systems, because that's what 90% of future market gardeners want. With this base, I was able to develop my on-farm educational program while restaurants were closed during the pandemic." He has now trained four market gardeners who have since started biointensive farms, three in Nantes and one in Lyon. For one year, students at the Chamber of Agriculture training center will also be able to test the biointensive model on a 2.5-acre plot. It's a first in the region. "I love passing on my method and showing people that it works," he says. About 750 square feet in my building are dedicated to hosting apprentices." Jean-Michel devotes half of his time to his market garden, and the other half is spent teaching, for a total of 40 hours per week. To maintain his educational plots, he has help from an apprentice and interns. "My gross revenue is €41,000 (US$44,500) from vegetable sales and teaching. Based on my calculations, that is roughly €20 per square meter (US$2 per square foot). I could increase my revenue if I worked on the market garden full-time. So this leaves me a lot of room for progress, which is quite encouraging."

In 2021, Jean-Michel's son, Simon, and his wife set up shop next door, where they started a biointensive flower farm. It seems Microferme des Anges is continuing to diversify its operations—as a family! Jean-Michel is even thinking of opening a small restaurant on his farm, where clients could enjoy his tasty vegetables on-site.

KEY DATES

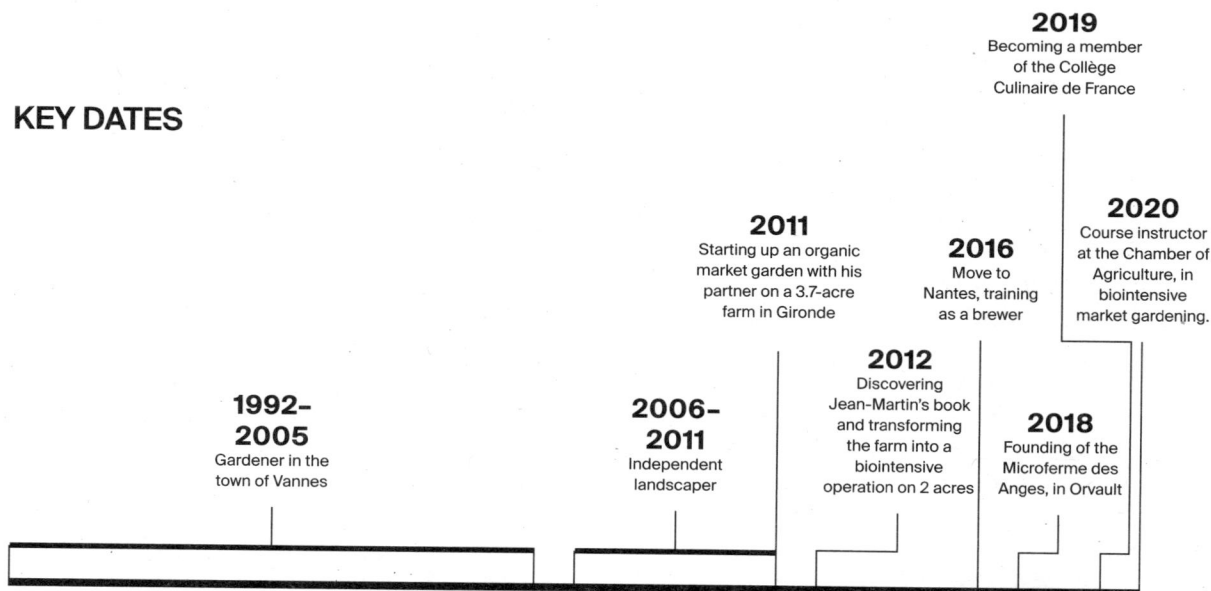

2019
Becoming a member of the Collège Culinaire de France

2020
Course instructor at the Chamber of Agriculture, in biointensive market gardening.

2011
Starting up an organic market garden with his partner on a 3.7-acre farm in Gironde

2016
Move to Nantes, training as a brewer

2012
Discovering Jean-Martin's book and transforming the farm into a biointensive operation on 2 acres

2018
Founding of the Microferme des Anges, in Orvault

1992–2005
Gardener in the town of Vannes

2006–2011
Independent landscaper

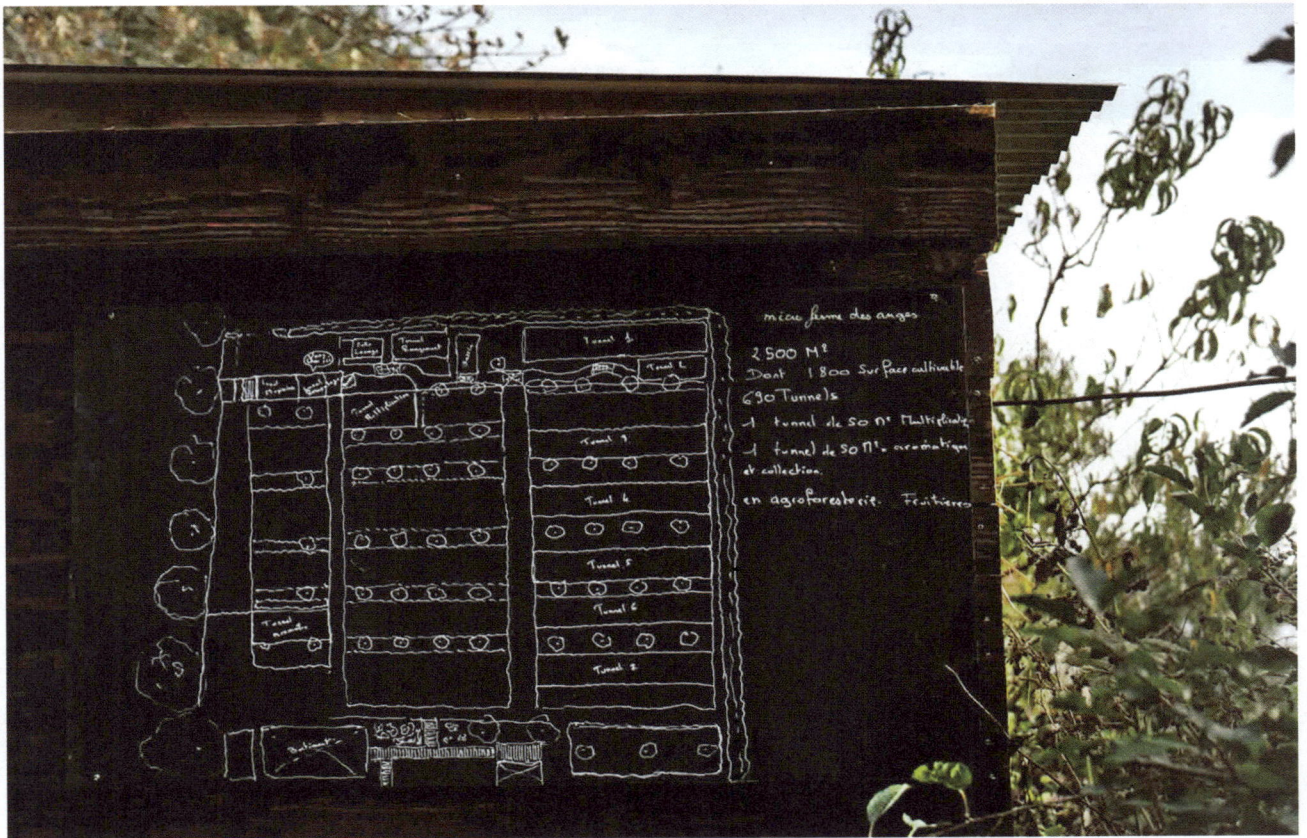

THE BIOINTENSIVE SYSTEM, ACCORDING TO JEAN-MICHEL

In my opinion, no small-scale farming model is more profitable than this one. It's the agricultural model of the future, and one of the answers to current land access challenges. I've tried all the methods—tractors, tillers, and now the tilther—and adapted tools to fit the size of my farm. With the biointensive model and its tools, you can work fewer hours, and have more time for your personal life.

Glossary

Agroecology: A holistic approach to the interaction between human beings and their environment, where special attention is paid to soil life and its role in agricultural production. It includes conservation agriculture, regenerative agriculture, organic agriculture, and agroforestry as well as practices that promote biodiversity (the use of hedges, ponds, grass strips, etc.) and crop diversification (heirloom varieties, diverse crop rotations, cover crops, etc.).

Agroforestry: Agroforestry seeks to combine trees with crops and livestock on the same plot of land. It provides many advantages: diversification, improving yields, restoring soil fertility through biomass inputs, improving water quality and retention, maintaining biodiversity, sequestering carbon, adding shade and moisture in periods of high heat, providing a windbreak, etc.

Aisles: Space between 2 permanent beds that allows growers to move around the farm.

Carbon/nitrogen ratio: This is the amount of nitrogen and carbon present in organic matter. If a material has a 25:1 carbon/nitrogen ratio, it means that it contains 25 times more carbon than nitrogen.

Caterpillar tunnel: Less expensive than greenhouses and smaller than classic tunnels, caterpillar tunnels can shelter crops year-round. These crops will be healthier and provide better yields than if they were grown without a shelter. Four people can easily move a tunnel to another set of beds in just a few hours.

Chambers of agriculture (France): In France, Chambers of agriculture operate in the departmental, regional, and federal branches of government. They represent all parties in the realm of agriculture, rural life, and forestry: farmers, landowners, employees, professional groups. Through the regional government, they support applicants looking to start a farm.

Collège Culinaire de France: Association founded by 15 renowned chefs in 2011. It has more than 3,000 members, restaurateurs and artisans, operating throughout the country and its territories. The organization embodies a new culture of culinary craftsmanship, presented as an "alternative to the industrialization and standardization of production, catering, and food."

Compost: The result of organic waste, combined with water and oxygen, being broken down by microorganisms. This process can occur in a pile or composter, by creating a balanced mixture of "fresh" green waste (vegetable peels, mowing and crop residues, etc.) and "dry" brown waste (dead leaves, cardboard boxes, etc.). By amending soil with compost, growers can increase its organic matter and mineral content.

Cover crops: Also known as green manures, these are sown in the fall or spring and act as a ground cover. They may be used as a protective mulch or mowed when flowering, then left to decompose quickly while releasing nutrients. They can also be turned in after mowing.
They improve the composition and structure of soil and generate biomass. In the soil, they concentrate nutrients that will be beneficial for the following crops. Mustard, rye, phacelia, and buckwheat cover crops develop an important root system that aerates the soil and draws minerals up from deep in the ground. They collect these nutrients and then release them when they decompose. In comparison, plants in the Fabaceae family (legume family), such as alfalfa, beans, lupine, clover, vetch, and peas, have developed a unique feeding system that brings nitrogen into the plant tissue. In fact, if you look at their roots, you'll see that they have small bulges, or nodules. Inside the nodules, bacteria live in complete symbiosis with the plant, feeding on the substances created in the leaves during photosynthesis. In exchange, they provide the plant with nitrogen drawn from the air, in a process known as "nitrogen fixing." And through this collaborative work, Fabaceae are the only plants capable of fixing atmospheric nitrogen.

Crop rotation: To bring new and essential nutrients into the soil, it's important to not repeat the same crop every year. The key is to grow a wide range of species! Using cover crops, intercropping, and vegetative cover can help regenerate the soil: choose the best solution based on the crop that you plan to grow next.

CSA boxes/baskets (AMAP system in France): The concept is simple: customers pay a farmer in advance to receive a weekly delivery of freshly harvested produce.
The AMAP program (Association pour le maintien d'une agriculture paysanne) functions like the CSA concept. The major difference is that the AMAP name is trademarked, and farmers using this model are asked to follow guidelines set out in a National Charter.

Dotation Jeune Agriculteur (DJA, Funding for young farmers): Financial support program with 80% funding from the European Union and 20% from the state. The amount granted depends on the location of the new farm as well as the nature and scale of the project. The goal of this support program is to help new farmers set up their operation under conditions that are conducive to the success of the project; hence, DJA requirements apply to the applicant and the viability of the project. To qualify, candidates must be under 40 years old and have the required education and experience (known as capacité agricole in France).

Fertilization plan: Allows users to track and plan soil inputs like nitrogen, phosphorus, and potassium, to improve yields. Based on results, the dosages can be tweaked from one season to the next.

Groupement Agricole d'Exploitation en Commun (GAECC—French collective farming program): Agricultural civil society that gives farmers an opportunity to pool resources and work collectively, under conditions that are similar to those seen on family farms.

Inputs: Products added to soil (in the ground or in containers). Examples include fertilizers, plant protection products, and pesticides.

Manure: A mix of green manure (straw, fodder, etc.) and livestock feces, used as a fertilizer.

Market gardening on living soil (MSV, Maraîchage sur Sol Vivant): MSV* was first developed by French market gardeners in 2012. It encompasses agroecological principles and practices that "put soil back at the heart of the crop systems by guaranteeing shelter and sustenance for macro and microfauna in the soil," to promote "carbon farming" that increases soil moisture while providing structure, reduces manure and irrigation requirements, protects the environment, and limits the amount of work required of market gardeners, in the long run. This approach is based on two main principles: maintaining a permanent ground cover (green manures, cover crops, mulches, plant diversification, tarps, etc.) and cutting down on tilling. It draws upon methods and philosophies like permaculture, biointensive market gardening, and natural farming.

Microfarming: The basic principles of microfarming are autonomy, small scale (to better manage the land with little to no mechanization), and the development of biodiversity.

Microgreens: Plants harvested at the very early shoot stage for consumption. They are quite rich in amino acids (digestible proteins), sugars, fatty acids, enzymes, vitamins, minerals, micronutrients, carotene, fiber, and chlorophyll. Radish, sunflower, watercress, arugula, broccoli, beet, mustard, and many others can be grown as microgreens.

Mulching: Mulching is a technique that consists of covering soil around the base crops with straw, hay, mowing residues, hemp hurd, compost, etc. It helps with weed management, preserves soil moisture, and protects the soil from erosion.

Organic matter: Organic matter is the high-carbon matter generally produced by living beings, plants, animals, and microorganisms. Unlike mineral content, organic matter is often biodegradable, and therefore, it can be easily recycled into compost or biogas. In addition to carbon and water, its essential components, it may also contain hydrogen, oxygen, nitrogen, phosphorus, sulfur, iron, and more. Organic matter that is more complex, hard, dry, woody, and brown tends to have more structure. When you can easily grab big handfuls of your compost, it likely has a high carbon ratio. In comparison, when the organic matter is more humid, green, flexible, and without much consistency, or even liquid, your compost will have a higher nitrogen content.

Organopónicos: Originally developed in Cuba, organopónicos are urban vegetable farms with beds that are 25 to 50 feet long and 12 to 24 inches wide. Beds are surrounded by a low wall, then filled in with soil and gradually amended with organic matter. Built to sit higher than ground level, they allow growers to farm on polluted land.

Permaculture: From the words "agriculture" and "permanent," "permaculture" was coined in the 1970s, by biologist Bill Mollison and agronomist David Holmgren, both Australian. It brings together principles and techniques for land and crop management that are both ancestral and innovative, under a broader concept: design. It aims to make spaces harmonious, productive, autonomous, and naturally regenerating ecosystems that respect nature and its inhabitants.

Permanent beds: Strips of land that are raised to roughly 4 to 6 inches above ground level, separated by pathways for traffic, and in which crops are planted. Farmers can choose the bed length that they deem most suitable for their operation. We recommend a bed length between 50 and 100 feet, with a 29-inch width, so that you can standardize materials and equipment like silage tarps, irrigation lines, insect netting, etc. By growing crops on standardized beds, you'll be able to optimize use of these materials, and thus have to purchase smaller quantities. So that you can walk between the beds with a wheelbarrow, we recommend 18-inch aisles. These dimensions allow workers to straddle or step over a bed without trampling it, to minimize soil compaction. Beds become the unit of measurement on a market garden (rather than acres) for tasks like calculating amendment doses, crop planning, etc. The beds can be grouped into blocks, with the same number of beds. At Ferme des Quatre-Temps, each block contains 10 beds.

Plow pan: Sometimes referred to as a tillage pan or a hardpan, this is a compact soil layer, often at the farm's tilling depth, that is about 1 inch thick. It tends to be caused by the plowshare pressing down on the soil.

Ramial chipped wood (RCW): Wood chips made from wood shoots, twigs, and branches with a diameter of less than 2¾ inches. Typically, this is waste from pruning deciduous trees. The wood is so fresh that it's practically still alive. It stimulates soil life, as it would in forest soils. This is why RCW should be viewed as an "aggrader" (able to counter degradation) rather than an amendment because it fosters the development of cryptogamic life in soils and promotes pedogenesis.

Regenerative agriculture: A set of agronomic practices aimed at restoring the quality of agricultural soils through sound farming methods. By introducing more life and biodiversity into soils,

* https://normandie.maraichagesolvivant.fr/lassociation/

regenerative agriculture promotes the development of organic matter, creating a richer, more fertile soil that naturally sequesters more carbon. It thus contributes to mitigating global warming. Examples of practices: fertilizing soil with organic matter such as compost or manure, reducing tillage, planting hedges around plots of land to provide shelter for wildlife and protect farmed land from erosion by wind and rain.

Retention basin or pond: Sometimes called stormwater management ponds, these underground or open-air basins are built to store stormwater. They are usually dug with an excavator and covered with a waterproof tarp.

Slow Flower: The Slow Flower movement began in the 2000s in the United States and advocates for the production and consumption of local and seasonal flowers.

Swales: Wide, shallow ditches that help collect and move rainwater, and improve water infiltration. This system features gently sloping sides that allow farmers to more effectively manage hydraulic flow and redirect overflow to a retention basin.

Tilther: A smaller version of the walk-behind tractor.

Walk-behind tractor: A motorized tool used to turn over topsoil.

About the Authors

Jean-Martin Fortier, a distinguished author, educator, and proponent of regenerative agriculture on a human scale, has been spearheading a movement in sustainable farming alongside his wife, Maude-Hélène Desroches. For over 20 years, their venture, Les Jardins de la Grelinette, an efficient and productive 2-acre microfarm, has thrived on the principles of biointensive agriculture.

In 2015, Jean-Martin established Ferme des Quatre-Temps, a research farm where he trains apprentices and develops new strategies for market gardening, including year-round production. His methods are taught worldwide through the Market Gardener Masterclass, an online course designed to empower and educate small-scale farmers. He is the founder of Growers & Co., a tool and apparel company that celebrates sustainable, local, and equitable food producers.

Jean-Martin is the co-author of *The Winter Market Gardener* and author of *The Market Gardener*, which has sold over 250,000 copies in nine languages. His contributions to organic farming and regenerative agriculture earned him the Meritorious Service Cross from the Governor General of Canada. He farms and lives in Quebec, Canada.

Aurélie Sécheret is a freelance author, journalist, and editor specializing in social and ecological issues with a particular interest in food security. A passionate urban farmer, she works with the noted grassroots French agricultural association Veni Verdi to develop local farming projects in the heart of busy city neighborhoods.

Create a future where humans live in harmony with nature and with each other

INSTITUT JARDINIER MARAICHER

themarketgardener.com

Founded by Jean-Martin Fortier, the Market Gardener Institute aims to inspire, educate, and support today's farmers throughout their journey, by providing them with the technical skills required to succeed in their valuable agricultural work.

We strongly believe that agriculture on a human scale has the potential to transform communities, and that the proliferation of ecological and regenerative microfarms around the world represents a real solution to global challenges and problems. Our mission is to invest in this transition.

Grower's Guides from the Market Gardener

Coming soon: *Root Vegetables, Living Soil, Fruit Vegetables, A Year of Vegetables*

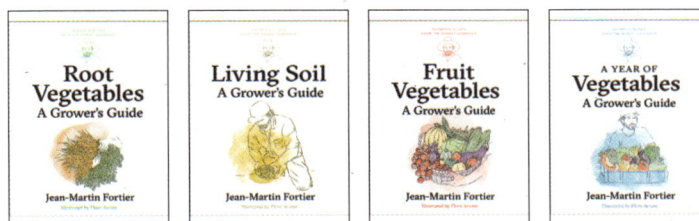

Resources and Bibliography

Planning Tools

Heirloom: a crop planning and farm management tool developed by the Market Gardener Institute (heirloom.ag)

Sales and Marketing Tools

To accept card purchases: https://squareup.com/us/en

References

Coleman, Eliot. 2018. *The New Organic Grower, 3rd Edition*. Chelsea Green Publishing (originally published in 1989).

Fortier, Jean-Martin. 2014. *The Market Gardener: A Successful Grower's Handbook for Small-Scale Organic Farming*. New Society Publishers.

Fortier, Jean-Martin and Catherine Sylvestre. 2023. *The Winter Market Gardener: A Successful Grower's Handbook for Year-Round Harvests*. Translated by Laurie Bennett. New Society Publishers.

Moreau, J.G. and J.J. Daverne. *Manuel pratique de la culture maraîchère de Paris*.1845, available for free and in French only, at: http://gallica.bnf.fr

Continuing Education

The Market Gardener Institute Masterclass. https//themarketgardener.com

ABOUT NEW SOCIETY PUBLISHERS

New Society Publishers is an activist, solutions-oriented publisher focused on publishing books to build a more just and sustainable future. Our books offer tips, tools, and insights from leading experts in a wide range of areas.

We're proud to hold to the highest environmental and social standards of any publisher in North America. When you buy New Society books, you are part of the solution!

At New Society Publishers, we care deeply about *what* we publish—but also about *how* we do business.

- This book is printed on 100% **post-consumer recycled paper**, processed chlorine-free, with low-VOC vegetable-based inks (since 2002)
- Our corporate structure is an innovative employee shareholder agreement, so we're one-third employee-owned (since 2015)
- We've created a Statement of Ethics (2021). The intent of this Statement is to act as a framework to guide our actions and facilitate feedback for continuous improvement of our work
- We're carbon-neutral (since 2006)
- We're certified as a B Corporation (since 2016)
- We're Signatories to the UN's Sustainable Development Goals (SDG) Publishers Compact (2020–2030, the Decade of Action)

To download our full catalog, sign up for our quarterly newsletter, and to learn more about New Society Publishers, please visit newsociety.com.

ENVIRONMENTAL BENEFITS STATEMENT

New Society Publishers saved the following resources by printing the pages of this book on chlorine free paper made with 100% post-consumer waste.

TREES	WATER	ENERGY	SOLID WASTE	GREENHOUSE GASES
37	3,000	16	120	16,000
FULLY GROWN	GALLONS	MILLION BTUs	POUNDS	POUNDS

Environmental impact estimates were made using the Environmental Paper Network Paper Calculator 4.0. For more information visit www.papercalculator.org

FSC
www.fsc.org
MIX
Paper | Supporting responsible forestry
FSC® C016245

SDG PUBLISHERS COMPACT

Certified B Corporation

new society PUBLISHERS
www.newsociety.com